Selected Papers from the 16th International Symposium on Magnetic Bearings (ISMB16)

Selected Papers from the 16th International Symposium on Magnetic Bearings (ISMB16)

Editors

**Jin Zhou
Huachun Wu
Satoshi Ueno
Feng Sun**

MDPI • Basel • Beijing • Wuhan • Barcelona • Belgrade • Manchester • Tokyo • Cluj • Tianjin

Editors
Jin Zhou
Nanjing University of
Aeronautics and Astronautics
China

Huachun Wu
Wuhan University of Technology
China

Satoshi Ueno
Ritsumeikan University
Japan

Feng Sun
Shenyang University of Technology
China

Editorial Office
MDPI
St. Alban-Anlage 66
4052 Basel, Switzerland

This is a reprint of articles from the Special Issue published online in the open access journal *Actuators* (ISSN 2076-0825) (available at: https://www.mdpi.com/journal/actuators/special_issues/ISMB16).

For citation purposes, cite each article independently as indicated on the article page online and as indicated below:

LastName, A.A.; LastName, B.B.; LastName, C.C. Article Title. *Journal Name* **Year**, *Article Number*, Page Range.

ISBN 978-3-03943-070-3 (Hbk)
ISBN 978-3-03943-071-0 (PDF)

© 2020 by the authors. Articles in this book are Open Access and distributed under the Creative Commons Attribution (CC BY) license, which allows users to download, copy and build upon published articles, as long as the author and publisher are properly credited, which ensures maximum dissemination and a wider impact of our publications.

The book as a whole is distributed by MDPI under the terms and conditions of the Creative Commons license CC BY-NC-ND.

Contents

About the Editors . . . . **vii**

Preface to "Selected Papers from the 16th International Symposium on Magnetic Bearings (ISMB16)" . . . . **ix**

Virginie Kluyskens, Joachim Van Verdeghem and Bruno Dehez
Experimental Investigations on Self-Bearing Motors with Combined Torque and Electrodynamic Bearing Windings
Reprinted from: *Actuators* **2019**, *8*, 48, doi:10.3390/act8020048 . . . . **1**

Dominik Wimmer, Markus Hutterer, Matthias Hofer and Manfred Schrödl
Space Vector Modulation Strategies for Self-Sensing Three-Phase Radial Active Magnetic Bearings
Reprinted from: *Actuators* **2019**, *8*, 41, doi:10.3390/act8020041 . . . . **23**

Daniel Franz, Maximilian Schneider, Michael Richter and Stephan Rinderknecht
Thermal Behavior of a Magnetically Levitated Spindle for Fatigue Testing of Fiber Reinforced Plastic
Reprinted from: *Actuators* **2019**, *8*, 37, doi:10.3390/act8020037 . . . . **41**

Josef Passenbrunner, Gerald Jungmayr and Wolfgang Amrhein
Design and Analysis of a 1D Actively Stabilized System with Viscoelastic Damping Support
Reprinted from: *Actuators* **2019**, *8*, 33, doi:10.3390/act8020033 . . . . **59**

Masahiro Osa, Toru Masuzawa, Ryoga Orihara and Eisuke Tatsumi
Performance Enhancement of a Magnetic System in a Ultra Compact 5-DOF-Controlled Self-Bearing Motor for a Rotary Pediatric Ventricular-Assist Device to Diminish Energy Input
Reprinted from: *Actuators* **2019**, *8*, 31, doi:10.3390/act8020031 . . . . **77**

Branimir Mrak, Bert Lenaerts, Walter Driesen and Wim Desmet
Optimal Magnetic Spring for Compliant Actuation—Validated Torque Density Benchmark
Reprinted from: *Actuators* **2019**, *8*, 18, doi:10.3390/act8010018 . . . . **91**

Alexander H. Pesch and Peter N. Scavelli
Condition Monitoring of Active Magnetic Bearings on the Internet of Things
Reprinted from: *Actuators* **2019**, *8*, 17, doi:10.3390/act8010017 . . . . **107**

Joachim Van Verdeghem, Virginie Kluyskens and Bruno Dehez
Stability and Performance Analysis of Electrodynamic Thrust Bearings
Reprinted from: *Actuators* **2019**, *8*, 11, doi:10.3390/act8010011 . . . . **121**

About the Editors

Jin Zhou (Professor) received her Ph.D. degree in mechanical engineering from the China University of Mining and Technology (CUMT) in 2001. From 2012 to 2013, she was a visiting scholar in the rotating machinery and control laboratory (ROMAC) of the University of Virginia. She is currently a full professor in the College of Mechanical and Electrical Engineering, NUAA. Her research focuses on magnetic bearings and vibration control. She was the member of the Program Committee of the 14th International Symposium on Magnetic Bearings (ISMB, 2014), and the Program Chair of the 16th International Symposium on Magnetic Bearings (ISMB, 2018). She was also an elected member of the International Advisory Committee of ISMB in 2018.

Huachun Wu (Professor), male, born in November 1976, received his Ph.D. degree from the Wuhan University of Technology in 2005. He is currently the deputy dean of the School of Mechanical and Electronic Engineering, at the Wuhan University of Technology, as well as the director of the Hubei Provincial Engineering Technology Research Center for Magnetic Suspension. His main research involves the design and characteristic analysis for the maglev system, controlling and fault diagnosis for magnetic bearing, vibration analysis and the testing of rotating machinery.

Satoshi Ueno (Professor) received his D. Eng. degrees from Ibaraki University, Hitachi, Japan, in 2000. He is currently a professor of the Department of Mechanical Engineering, Ritsumeikan University, Japan. His research interests include active magnetic bearings, self-bearing motors, and their applications. Prof. Ueno is a member of IEEE, the Japan Society of Mechanical Engineers, and several other societies.

Feng Sun is a professor in the School of Mechanical Engineering at Shenyang University of Technology, China, and a guest professor of Kochi University of Technology, Japan. He received his Ph.D. from the Kochi University of Technology, Japan, in 2010. His main research concerns the multi-type actuators and the control technology of the mechanical system. He has applied the magnetic suspension technology and piezoelectric actuators to the electric spindle, laser cutting, electrical discharge machining, and automotive intelligent suspension. He has published more than 130 academic papers, two books, one monograph, and the monograph was funded by the National Fund for the Publication of Scientific and Technological Academic Works. He holds 21 authorized invention patents.

Preface to "Selected Papers from the 16th International Symposium on Magnetic Bearings (ISMB16)"

Magnetic bearing is an electromagnetic device that provides magnetic force in order to suspend shafts, in contrast to other conventional bearings which rely on mechanical force. With the inherent distinguished features including the absence of mechanical wear, the elimination of lubrication, long life expectation, tunable stiffness and damping, as well as high attainable rotating speeds, magnetic bearings are widely applied in compressors, bearingless motors, and other high-speed rotating machinery applications. The fast and continued expansion of magnetic bearings has trigged enormous interest in both academia and industry. In 2018, the 16th International Symposium on Magnetic Bearings (ISMB16) was held in Beijing, China. In this symposium, six plenary speeches and 112 selected papers were presented. Herein, eight distinguished papers were selected to be published as Special Issues. These papers are suggested to provide new insights for the development of magnetic bearings, with emphasis on thermal behavior analysis, electrodynamic thrust bearings analysis, viscoelastic damping design, magnetic spring, condition monitoring, self-bearing motors performance.

Jin Zhou, Huachun Wu, Satoshi Ueno, Feng Sun
Editors

Article

Experimental Investigations on Self-Bearing Motors with Combined Torque and Electrodynamic Bearing Windings †

Virginie Kluyskens, Joachim Van Verdeghem and Bruno Dehez *

Center for Research in Mechatronics (CEREM), Institute of Mechanics, Materials and Civil Engineering (IMMC), Université catholique de Louvain (UCL), 1348 Louvain-la-Neuve, Belgium;
virginie.kluyskens@uclouvain.be (V.K.); joachim.vanverdeghem@uclouvain.be (J.V.V.)
* Correspondence: bruno.dehez@uclouvain.be
† This paper is an extended version of our paper published in Kluyskens, V.; Van Verdeghem, J.; Dumont C.; Dehez, B. Experimental Investigations on a Heteropolar Electrodynamic Bearing-self-bearing Motor. In Proceedings of the 16th International Symposium on Magnetic Bearings (ISMB), Beijing, China, 13–17 August 2018.

Received: 7 May 2019; Accepted: 5 June 2019; Published: 11 June 2019

Abstract: The centering guidance forces in self-bearing permanent magnet motors are magnetically integrated with the torque generation windings, and can take place in a single multifunction winding. This radial guidance is usually actively controlled as a function of the rotor position, with the drawbacks associated to actively controlled devices. This article describes how multifunction windings can passively generate electrodynamic centering forces without the need for specific additional electronics, and simultaneously a driving torque if fed by a power supply. It shows the experimental electromotive force (EMF) measures, both for the electrodynamic centering and for the motor functions, obtained on a prototype, operating in quasistatic conditions. It also shows the measured radial forces generated by the electrodynamic bearing and the measured drive torque in these conditions. These measures show a good agreement with model predictions. These measures also confirm the theoretical conclusions stating that it is possible to generate passive guidance forces and torque simultaneously in a single winding. The effect of adding external inductors on the coils of the prototype is also investigated by experimental measures and model predictions on the bearing radial forces, and on the motor driving torque. It is shown that these external inductors mainly affect the radial guidance forces with minor impact on the torque.

Keywords: self-bearing motor; electrodynamic bearing; passive levitation

1. Introduction

Self-bearing motors are an attractive solution to issues related to compactness and maximum spin speed. Various kinds of electric machines have been studied for self-bearing operation, where self-bearing operation means a system in which the drive function and the bearing function are magnetically integrated. This bearing function is always achieved, at least for one degree of freedom, by the generation of guidance forces controlled by modulating a current as a function of the rotor position. Examples of various types of self-bearing motors can be found in the literature, for example, in [1–5]. Active guidance in self-bearing-motors can be related to active magnetic bearings (AMBs), which allow reaching relatively high stiffness values, high positioning precision, and have reached a certain level of industrial maturity. However, the complexity, cost, and overall dimensions associated with this control system can be prohibitive, e.g., for low-rated power applications.

Passive guidance for self-bearing motors has not been substantially investigated. However, some references show that when short-circuiting the guidance windings of self-bearing motors, some restoring forces appear. Reference [6] shows a self-bearing induction motor with the rotor mounted on a flexible shaft and supported by external bearings. The bearing windings, usually controlled to provide an active guidance for the rotor, were simply short-circuited. The article shows that the vibrations decreased with short-circuited bearing windings compared with open circuit bearing windings, but less than when the bearing windings are actively controlled. The same kind of observation was reported in [7] for a two-pole induction motor. Finally, the principle of radial forces generation with bearing windings in short circuit for a self-bearing switched reluctance machine is studied in [8] and a reduction of the vibrations was observed.

This guidance based on short-circuited coils could be improved if the short-circuited windings were specially designed and optimized for passive guidance. This is the case of electrodynamic magnetic bearings (EDBs). EDBs are based on forces resulting from the interaction between a magnetic field and currents induced in conductors by a variation of the magnetic field seen by these conductors. This variation arises from a space variation of the field and a relative motion between the conductor and the magnetic field. Preferably, electrodynamic bearings are designed in such a way that currents are only induced when the rotor is off-centered, in order to avoid unnecessary losses in the centered position, these latter are null flux electrodynamic bearings. However, these bearings are difficult to design as their stability depends on the rotor spin speed and on the damping present in the system [9]. Moreover, their stiffness depends on the spin speed, and the specific load capacity of EDBs remains lower than that of active magnetic bearings (AMBs) [10].

Centering homopolar EDBs have received much interest, resulting in dynamic models [9,11], prototypes, and a successful levitation test [12]. More recently, centering heteropolar bearings have been studied, leading to design rules on the windings so as to be null-flux [13]. Design rules are also given in [1] for dual-purpose windings (active radial guidance and motor windings) in self-bearing motors with no-voltage when the rotor is centered, which relates to null-flux windings in electrodynamic bearings. Regarding the dynamic behavior and the passive electrodynamic guidance forces generated in heteropolar EDBs, a model is provided in [14]. Heteropolar EDBs present a structure close to permanent magnet (PM) motors, and integration of the EDB inside the PM motor gives the opportunity to take advantage of the magnetic field produced by the permanent magnets already present inside the motor for the generation of radial forces. It can also be noted that PM motors are particularly well suited for high spin speed operation and that EDBs produce a maximal stiffness at high spin speed, which gives even more sense to this integration. This kind of integration, a centering heteropolar EDB inside a PM motor, has already been described in [15] for two cases: the first one with two distinct windings systems for the motor and the guidance functions, and the second one with one single multifunction winding system. A finite element (FE) study of an application case is also considered in this paper, showing the theoretical feasibility of such a device, for a slotless winding configuration.

The goal of this paper is to go a step further in the study of self-bearing PM motors, in which the passive electrodynamic radial guidance generation takes place in the same winding as the torque generation of the motor (one single multifunction winding). A prototype of a heteropolar centering EDB with a radial magnetic field has been constructed, and its guidance performances have been studied in [16]. Consecutive to the theoretical studies presented in [15], the prototype presented in [16] is experimentally investigated as a self-bearing motor in [17], to confirm its ability to also develop a driving torque. This article first outlines the experimental results obtained on the prototype without modifications [17], as well on the generated electromotive forces for the bearing function and for the motor function, as on the measures of the guidance forces and driving torque developed. These measures show that, in the range of spin speeds considered for the experimental measures, the prototype can indeed develop a driving torque and develop radial forces. However, those measures on the radial forces show that the restoring radial forces are smaller than the parasitic radial force. In this article, new experimental measures on the prototype are presented: additional external inductors are connected to the coils, which changes the electrical pole of the prototype. These new measures show that by changing the electrical pole of the prototype, it is possible to have a radial restoring force more important than the parasitic force within the same range of spin speeds. The influence of these additional external inductors on the value of the drive torque is also investigated. Finally, this article compares all the experimental measures, for the guiding forces and for the driving torque, with and without additional inductors, to the theoretical model predictions.

The present article is structured as follows: it first briefly explains how a single winding can be designed to act as an EDB winding and as a motor winding. In the next section, the model is briefly explained. Next, the prototype and the test bench are described. Next, the article shows quasistatic experimental results, in terms of electromotive force (EMF), both for the EDB and for the motor functions. Finally, the article shows the experimental measures obtained for the radial forces and driving torque when a load is connected to the prototype. These measures are shown for the prototype as it is, and when additional inductors are connected to the coils. Measures of the currents inside the coils are also shown. These measures are also compared to model predictions, before the conclusions.

2. Operating Principle

This section briefly describes the general operating principle of a single multifunction winding performing both drive and EDB principles, in the case of a PM rotor with one pole pair. In this case, $p = 1$, the motor winding also has one pole pair in order to achieve optimal magnetic coupling and to generate a driving torque. An example of such a winding, with a window frame configuration and for one phase, is shown in Figure 1a. Concerning the EDB winding, as explained in [6], when the rotor is an internal rotor, the EDB short-circuited winding has to have two pole pairs ($p + 1$) to be magnetically coupled to the harmonics linked to the rotor off-centering. An example of such a winding, also with a window frame configuration and for one phase, is shown in Figure 1b. From these figures, it can be seen that these two windings can be combined into one single multifunction winding. This is shown in Figure 2, consistently for one phase. This multifunction winding can be fed through terminals R_S and R_F to produce a torque: both coils are then connected in an antiparallel configuration. However, this winding also allows for a short circuit path consisting of both coils connected in series. The currents induced when the rotor is moving out-centered can then circulate in this short-circuit path and generate electrodynamic restoring forces.

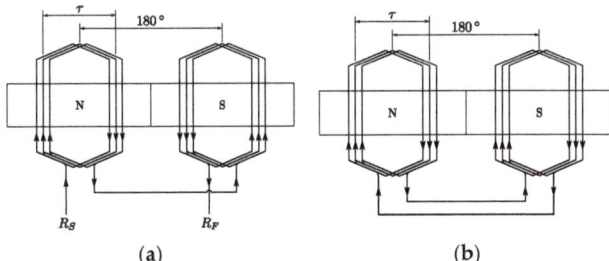

Figure 1. Unrolled view of a one-pole pair permanent magnet rotor with one phase (**a**) of a motor winding and (**b**) of an electrodynamic bearing winding if the rotor is internal.

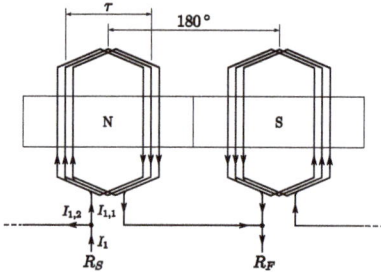

Figure 2. Unrolled view of phase 1 of a multifunction winding performing both motor and passive electrodynamic magnetic bearing (EDB) centering functions for a one-pole pair internal permanent magnet rotor.

3. Simplified Analytical Model

We make the following assumptions:

- the materials have linear magnetic characteristics,
- only the highest harmonics in the permanent magnet field distribution are considered,
- the rotor only undergoes translational eccentricity, i.e., the magnetic axis of the rotor and of the winding remain parallel,
- the rotor spin speed ω is constant.

Based on these assumptions, the magnetic vector potential due to the permanent magnet rotor has only one axial component, A_{Mz}, which can be derived as described in [6,7]. Finally, when the rotor is internal and has one pole pair, the magnetic vector potential is worth at a point P placed at coordinates (r, θ) from the stator center (as shown in Figure 3):

$$\vec{A} = A_{Mz}(r, \theta)\vec{e_z} = \left[A(r)\sin(\theta - \omega t) + \epsilon A^i(r)\sin(2\theta - \omega t - \phi)\right]\vec{e_z}, \tag{1}$$

where $A(r)$ and $A^i(r)$ are the amplitude of the vector potential component of periodicity $p = 1$ when the rotor is centered and of periodicity $p + 1 = 2$ when the rotor is off-centered, respectively. They depend on the geometric and magnetic properties of the system. The distance between the rotor and stator center is named ϵ, and the direction of off-centering is named Φ.

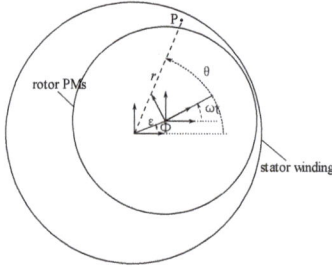

Figure 3. Frames and coordinates used for the model.

Considering window frame windings connected as illustrated in Figure 2, the magnetic flux seen by each coil of the winding of phase k can be calculated by integrating $\Phi = \oint_k \vec{A} \vec{dl}$, along the conductors of these coils, with $\vec{A}$ described in (1). From there, the electromotive forces $e(t)$ generated on each coil is calculated by derivation of the flux, as detailed in [8], resulting in:

$$e_0(t) = \omega K_{\Phi,mot} \sin(\delta_k - \omega t), \tag{2}$$

$$e_d(t) = \varepsilon \omega K_{\Phi,EDB} \sin(2\delta_k - \omega t - \phi), \tag{3}$$

where δ_k represents the angular position of the magnetic axis of the first coil of the winding of phase k. The first EMF (2), with phasor $\overline{E_0}$, is generated by the inductor in the windings when the rotor is centered, is independent of the out-centering amplitude and can be linked to the motor behavior of the winding. The flux constant $K_{\Phi,mot}$ is equal to $2A(R_{winding})\sin(\frac{\tau}{2})$. The second EMF (3), with phasor $\overline{E_d}$, is the EMF induced by the harmonics of the magnetic field appearing when the rotor is not centered. It is proportional to the decentering amplitude, and relates to the EDB behavior of the winding. Its flux constant $K_{\Phi,EDB}$ is equal to $2A^i(R_{winding})\sin(\tau)$. From these equations, it can be seen that there will be a compromise to make on the coil pitch τ: a coil pitch of 180° maximizes the motor behavior, but cancels the EDB behavior. A coil pitch of 90° maximizes the EDB behavior, but at the expense of a smaller flux constant related to the motor behavior.

Equivalent electrical circuits for each phase of the multifunction winding can be represented as in Figure 4 [10]. In this circuit, the left and right branches correspond respectively to the coils facing a north or a south pole inside one phase winding. For the case represented in Figure 2 ($p = 1$), the left branch simply corresponds to the left coil, and the right branch to the right coil. Both branches have the same resistance R, and same cyclic inductance L_c, as each coil inside the winding is identical. Two EMFs appear on each part of the circuit. The motor EMF, E_0, has the same sign on both coils of the winding, given the antiparallel connection between them. The bearing EMF, named E_d, has opposite signs for each coil given the series connection of the coils seen by the winding in this case.

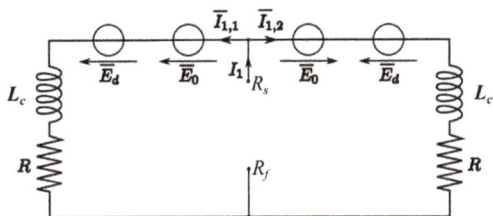

Figure 4. Equivalent electrical circuit of phase 1 of a multifunction winding for a one-pole pair internal inductor, with R the resistance of one coil, L_c, the cyclic inductance of one coil, $\overline{E_0}$, the electromotive force (EMF) when the rotor is centered and $\overline{E_d}$, the EMF linked to the center shift.

The electrical equations of the equivalent circuit of phase 1, as shown in Figure 4, are:

$$\overline{U}_{R_S-R_F} + \overline{E}_0 + \overline{E}_d - R\overline{I}_{1,1} - j\omega L_c \overline{I}_{1,1} = 0, \tag{4}$$

$$\overline{U}_{R_S-R_F} + \overline{E}_0 - \overline{E}_d - R\overline{I}_{1,2} - j\omega L_c \overline{I}_{1,2} = 0, \tag{5}$$

$$\overline{I}_{1,1} + \overline{I}_{1,2} = \overline{I}_1. \tag{6}$$

The same equations can be written for the N phases of the system.

In Equation (6), the total current $\overline{I}_1$ refers to the current supplied by the power supply, and it only contributes to the torque generation. Indeed, when the rotor is centered, the induced EMF $\overline{E}_d$ is equal to zero. The current distribution in each loop is then balanced and equates to:

$$\frac{1}{2}\overline{I}_1 = \overline{I}_{1,1} = \overline{I}_{1,2} = \frac{\overline{U}_{R_S-R_F} + \overline{E}_0}{R + j\omega L_c} \tag{7}$$

When the rotor is out-centered, an unbalance between the currents in each loop appears, generated by the EMF E_d. However, the total current I_1 remains at the same value as when the rotor is centered:

$$\overline{I}_{1,1} = \frac{\overline{U}_{R_S-R_F} + \overline{E}_0}{R + j\omega L_c} + \frac{\overline{E}_d}{R + j\omega L_c}, \tag{8}$$

$$\overline{I}_{1,2} = \frac{\overline{U}_{R_S-R_F} + \overline{E}_0}{R + j\omega L_c} - \frac{\overline{E}_d}{R + j\omega L_c}, \tag{9}$$

$$\overline{I}_1 = 2\frac{\overline{U}_{R_S-R_F} + \overline{E}_0}{R + j\omega L_c}. \tag{10}$$

The bearing restoring forces can be predicted by the state-space model presented in [9], which links the two degrees of freedom point mass rotor dynamic behavior to the electrodynamic forces developed inside a heteropolar bearing.

$$\begin{bmatrix} \dot{F}_x \\ \dot{F}_y \\ \ddot{x} \\ \ddot{y} \\ \dot{x} \\ \dot{y} \end{bmatrix} = \begin{bmatrix} -\frac{R_{bearing}}{L_{c,bearing}} & \omega_e & -K_d - \frac{K_{\Phi,bearing}^2 N}{2L_{c,bearing}} & 0 & -\frac{R_{bearing} K_d}{L_{c,bearing}} & \omega_e\left(\frac{K_{\Phi,bearing}^2 N}{2L_{c,bearing}} + K_d\right) \\ -\omega_e & -\frac{R_{bearing}}{L_{c,bearing}} & 0 & -K_d - \frac{K_{\Phi,bearing}^2 N}{2L_{c,bearing}} & -\omega_e\left(\frac{K_{\Phi,bearing}^2 N}{2L_{c,bearing}} + K_d\right) & -\frac{R_{bearing} K_d}{L_{c,bearing}} \\ \frac{1}{M} & 0 & -\frac{C}{M} & 0 & 0 & 0 \\ 0 & \frac{1}{M} & 0 & -\frac{C}{M} & 0 & 0 \\ 0 & 0 & 1 & 0 & 0 & 0 \\ 0 & 0 & 0 & 1 & 0 & 0 \end{bmatrix} \begin{bmatrix} F_x \\ F_y \\ \dot{x} \\ \dot{y} \\ x \\ y \end{bmatrix} \tag{11}$$

In the system analyzed in this paper, the electrodynamic bearing function is constituted by two identical coils in series, which means that from a bearing point of view, the winding total inductance and resistance are $2L_c$ and $2R$, and its total flux constant is $2K_{\Phi,EDB}$. Considering a system without ferromagnetic yoke (no detent forces), the detent stiffness K_d is zero. When the rotor is a one-pole pair internal rotor, the model parameter ω_e representing the electric pulsation of the rotor is equal to ω. The equation governing the dynamics of the system then becomes:

$$\begin{bmatrix} \dot{F}_x \\ \dot{F}_y \\ \ddot{x} \\ \ddot{y} \\ \dot{x} \\ \dot{y} \end{bmatrix} = \begin{bmatrix} -\frac{R}{L_c} & \omega & -\frac{2K_{\Phi,EDB}^2 N}{L_c} & 0 & 0 & \omega\left(\frac{2K_{\Phi,EDB}^2 N}{L_c}\right) \\ -\omega & -\frac{R}{L_c} & 0 & -\frac{2K_{\Phi,EDB}^2 N}{L_c} & -\omega\left(\frac{2K_{\Phi,EDB}^2 N}{L_c}\right) & 0 \\ \frac{1}{M} & 0 & -\frac{C}{M} & 0 & 0 & 0 \\ 0 & \frac{1}{M} & 0 & -\frac{C}{M} & 0 & 0 \\ 0 & 0 & 1 & 0 & 0 & 0 \\ 0 & 0 & 0 & 1 & 0 & 0 \end{bmatrix} \begin{bmatrix} F_x \\ F_y \\ \dot{x} \\ \dot{y} \\ x \\ y \end{bmatrix} \tag{12}$$

In this equation, C represents the additional damping in the system, and M is the rotor mass. In quasistatic conditions, i.e., when the rotor is spinning in a fixed out-centered position, these equations can be simplified, and two force components appear: one restoring component, acting to re-center the rotor, and one parasitic component, acting perpendicular to the center shift. In this case, these two components are equal to:

$$F_{paral} = \frac{2N(K_{\Phi,EDB}\omega)^2 L_c}{R^2 + (\omega L_c)^2} \varepsilon, \tag{13}$$

$$F_{perp} = \frac{2N(K_{\Phi,EDB})^2 \omega R}{R^2 + (\omega L_c)^2} \varepsilon. \tag{14}$$

The electrodynamic drag torque produced by the bearing function can be calculated as follows:

$$T_{EDB} = -xF_y + yF_x. \tag{15}$$

4. Prototype and Test Bench

A prototype of a centering heteropolar electrodynamic bearing has been constructed [16] and is used in this article to investigate its passive self-bearing capacities, to eventually confirm the theoretical conclusions of [15], stating that in a slotless configuration, it is possible to simultaneously generate sufficient passive electrodynamic centering forces and a driving torque in a single multifunction winding.

This prototype consists of a one-pole pair permanent magnet rotor, and a three-phase multifunction winding. The one-pole pair permanent magnet is formed by an annular NdFeB permanent magnet with parallel polarization, mounted on the shaft, and generating a sinusoidal magnetic field inside the airgap. Each phase of the stator coils consists of two concentrated window frame coils at 180°. There is no ferromagnetic yoke behind those coils, and the nominal airgap when the rotor is centered is 4 mm. A picture of the prototype is shown in Figure 5, and its characteristics are presented in Table 1.

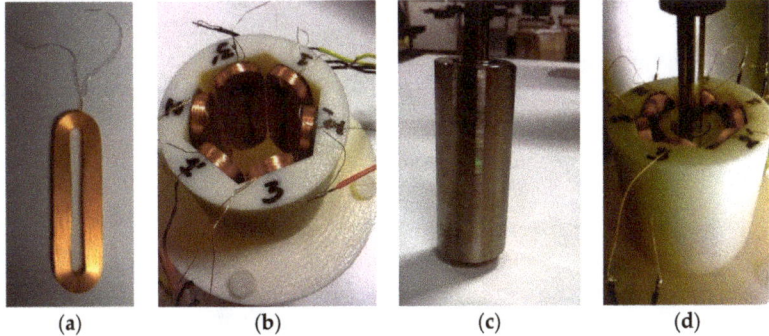

Figure 5. Picture of (**a**) One stator coil, (**b**) the stator, (**c**), the rotor, and (**d**) the centering heteropolar electrodynamic bearing prototype, investigated for its torque generation and self-centering capacities.

Table 1. Characteristics of the prototype.

Parameter	Value
Outer PM rotor diameter	25 mm
Inner PM rotor diameter	15 mm
Height of PM rotor	50 mm
Rotor magnet	NdFeB
PM remanence	1.3 T
Rotor shaft	Steel
Coil turns	560
Number of coils/phase	2
Number of phases	3
Coil wire diameter	0.2 mm
Coil total height	62 mm
Coil total width	17 mm
Coil thickness	4 mm
Stator inner diameter	33 mm
Nominal airgap	4 mm

A sectional schematic representation of the prototype is given in Figure 6, showing the current directions inside the windings when desiring a centering force or a torque, in accordance with the operating principle explained in the previous section.

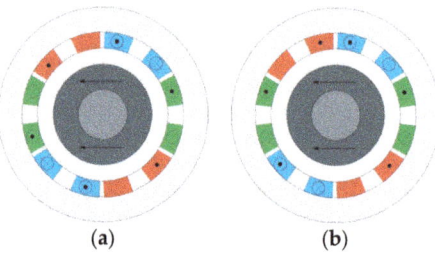

Figure 6. Schematic view of the prototype, with the current directions inside each phase resulting in (**a**) an electrodynamic centering force and (**b**) a motor torque.

The test bench is shown in Figure 7. It is designed to operate in quasistatic conditions, i.e., the rotor spins in a fixed out-centered position relatively to the stator. The rotor is driven by an external motor and its radial position is fixed. The stator is mounted on an xy manual stage, allowing displacing the stator with respect to the rotor with a micrometric precision. This test bench configuration allows to get rid of dynamic issues as the rotor is fixed by mechanical bearings. The relative displacement between the stator and the rotor is obtained by adjusting the stator position through the manual stage. The test bench is also equipped with a six-axis force sensor, measuring the reaction forces and torques on the stator winding. Finally, the prototype is encased inside an enclosure for safety.

Figure 7. Picture of the test bench for operation of the prototype in quasistatic conditions, with an external motor to drive the rotor.

5. Electromotive Forces

5.1. Experimental Results

An initial experiment was carried out to characterize the electromotive forces on the windings. The windings are left in open circuit, and the induced electromotive forces are measured on the two coils of each of the three phases ($m_1(t)$ and $m_2(t)$), which correspond to the phasors $\overline{M}_1$ and $\overline{M}_2$ on the equivalent circuit, in Figure 8, for one phase.

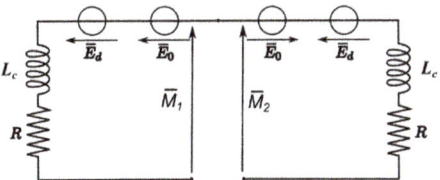

Figure 8. Principle of experimental measurements of the EMF shown on equivalent electrical circuit.

The induced EMFs are measured for various center shifts and various spin speeds: on the three phases, for spin speeds of 2400, 3600, 4800, and 6000 rpm, while displacing the rotor in the x–y plane by steps of 0.25 mm, with some additional 0.1 mm steps around the rotor center. A typical example of the obtained measures is shown in Figure 9a: within one phase, each coil presents a signal of the same frequency but with a difference in amplitude, depending on the rotor center shift. Fourier analysis of these signals is shown in Figure 9b and confirms the same frequency of the two signals. It can also be observed that the signal is almost purely sinusoidal, although a very small third harmonic is present.

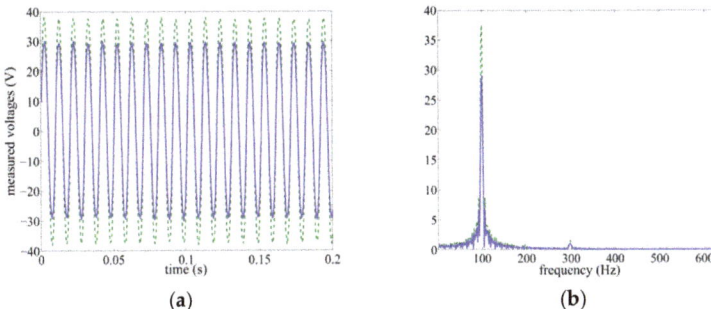

Figure 9. (a) Measured voltages and (b) Fourier analysis of measured voltages on phase 1 for a spin speed of 6000 rpm and a rotor center shift of 1 mm (along the y-axis).

This signal is postprocessed to extract the first harmonic and corresponding phase of each measure. The amplitude of the first harmonic of these measures is illustrated in Figure 10, for one phase, for different center shifts and spin speeds.

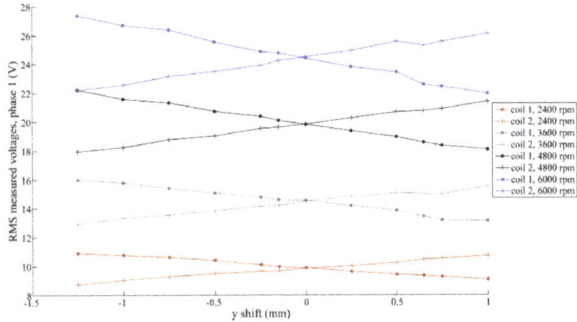

Figure 10. Root Mean square (RMS) value of first harmonic measured induced voltages at the coil terminals of phase 1, as a function of a rotor displacement along the y-axis, while centered along the x-axis, for rotor spin speeds of 2400, 3600, 4800, and 6000 rpm.

From these two measures obtained for each phase, the amplitudes of the electromotive force E_d, linked to the EDB, and the electromotive force E_0, linked to the motor can be deduced:

$$|\overline{E_0}| = |0.5(\overline{M_1} + \overline{M_2})|, \tag{16}$$

$$|\overline{E_d}| = |0.5(\overline{M_1} - \overline{M_2})|. \tag{17}$$

The results for the bearing EMF E_d and the motor EMF E_0 amplitudes deduced from the measurements are shown in Figure 11. It can be observed in Figures 11a and 12 that the motor EMF only depends on the spin speed, but not on the position of the center of the rotor. The results of the measures are very similar for each phase of the prototype. In Figure 11b, it can be observed that the bearing EMF increases proportionally with the center shift, and the slope increases with the spin speed. The bearing EMF measures also show (Figure 13) that due to geometric and magnetic imperfections, the geometric center of the system does not exactly correspond to its magnetic center. In addition, the bearing EMF should be zero when the rotor is perfectly magnetically centered, and the three phases should give the same results. However, we observe that the EMF reaches a minimum value which is situated on slightly different x points for each phase (Figure 13a). Similar observations can be made in the y-direction in Figure 13b. This illustrates that there is a slight dissymmetry between the phases. By displacing the rotor center by steps of 0.1 mm, the exact magnetic center is not found and the bearing EMF minimum value is slightly higher than zero: 0.2 V for the highest value at 6000 rpm. It can also be seen in Figure 13 that the dissymmetry between the phases remains small for all the measured points.

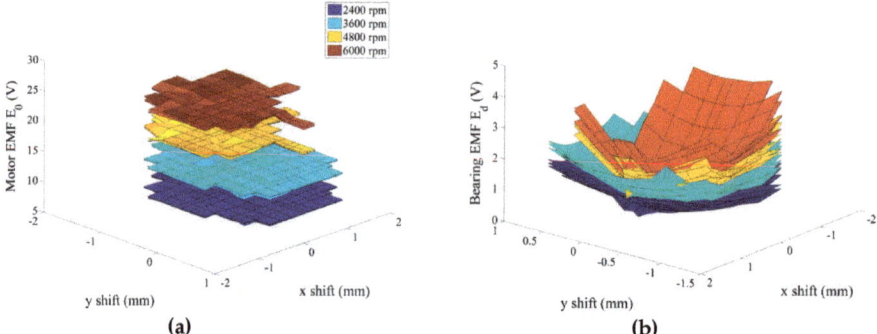

Figure 11. RMS values of (**a**) motor electromotive force E0 and (**b**) bearing electromotive force Ed as a function of the rotor position in the x–y plane for spin speeds of 2400, 3600, 4800, and 6000 rpm.

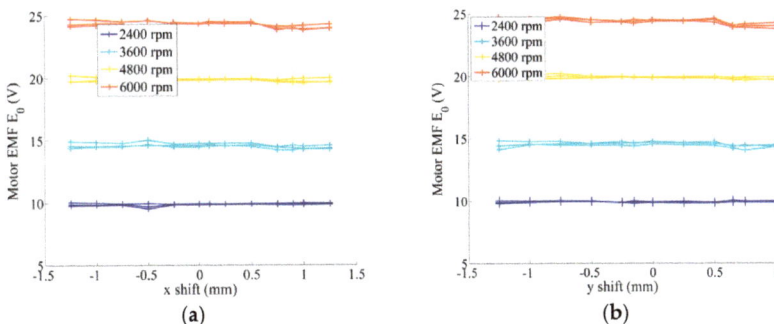

Figure 12. RMS values of motor electromotive force E_0 for each phase as a function of the rotor position (**a**) in the x axis and (**b**) in the y axis for spin speeds of 2400, 3600, 4800, and 6000 rpm.

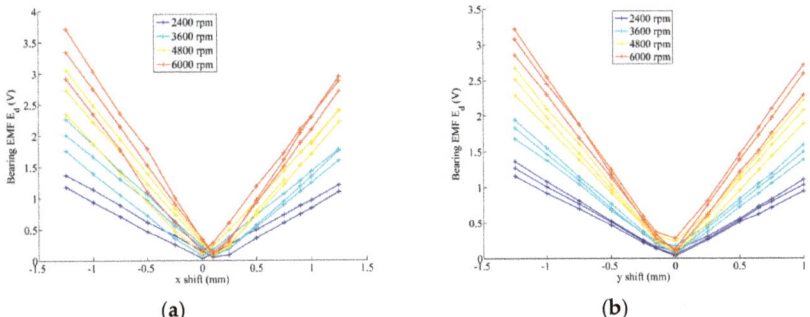

Figure 13. RMS values of bearing electromotive force E_d for each phase as a function of the rotor position (**a**) in the x axis and (**b**) in the y axis for spin speeds of 2400, 3600, 4800, and 6000 rpm.

5.2. EMF Experimental Model Comparisons

As shown in Section 3, the model predicts a bearing EMF proportional to the spin speed and to the center shift (3), and a motor EMF only proportional to the spin speed (2). This is indeed observed in Figure 11, Figure 12, and Figure 13. The flux constants $K_{\Phi,mot}$ and $K_{\Phi,EDB}$ in Equations (2) and (3) can be identified by static FE simulations or by performing curve fitting on the experimental measures, using a least square criterion. To identify those flux constants by FE, a 3D magnetostatic model is constructed as illustrated in Figure 14. The flux due to the permanent magnet rotor intercepted by each coil is calculated by integrating $\oint_{coil} \vec{A} d\vec{l}$. This is done for one off-centered rotor position and for angular positions of the rotor from 0 to 2π. The flux constants are then calculated by summing ($K_{\Phi,mot}$) or subtracting ($K_{\Phi,EDB}$) the flux values of the two coils of each phase and multiplying by the number of coil turns. For $K_{\Phi,EDB}$, this subtraction has to be divided by the rotor center shift. The numerical values obtained by FE and by curve fitting for these flux constants are given in Table 2.

Table 2. EMF model parameter identification.

Parameter	FE Identification	Exp. Curve Fitting	Relative Difference
$K_{\Phi,mot}$	0.0473	0.0389	21.6%
$K_{\Phi,EDB}$	4.8442	4.1366	14.6%

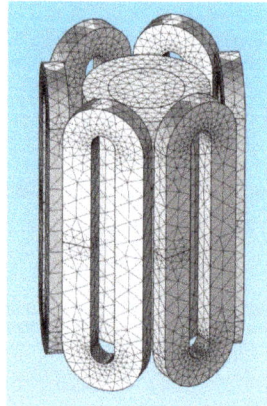

Figure 14. Mesh of the 3D Finite Element (FE) magnetostatic model of the prototype.

The bearing flux constant is similar when identified by FE simulations or by curve fitting; however, the motor flux constant is less similar between the two techniques. The measures for each phase and the model predictions, with the parameters identified by curve fitting, are represented in Figure 15, and show good correspondence.

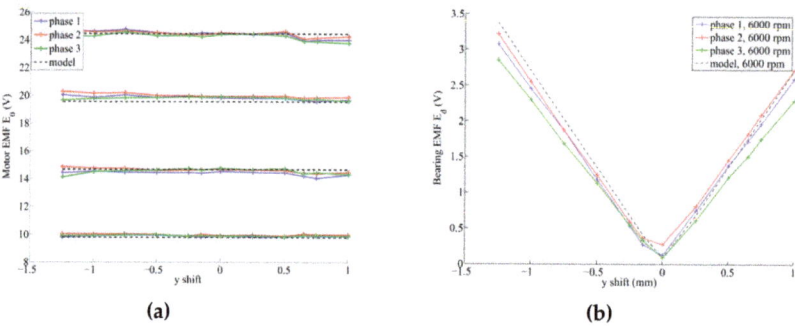

Figure 15. Experimental and model predictions for the three phases (**a**) of the motor electromotive force E_0 at spin speeds of 2400, 3600, 4800, and 6000 rpm and (**b**) of the bearing electromotive force E_d, at a spin speed of 6000 rpm as a function of the rotor position in the y-axis.

6. Force and Torque

A second experiment was carried out by connecting the two coils of each phase in series, and by driving the rotor by the external motor. A first set of measures on the restoring radial forces and the drag torque was obtained by leaving the motor connections (Figure 16a) of each phase in open circuit, while the prototype is only operating in bearing mode. To investigate the bearing and motor modes, a second set of measures is done by connecting a resistive charge on the motor connections $\{R_S, R_F\}$ and by measuring the radial forces and the resistive torque again (Figure 16b). As the torque generation abilities of the prototype are investigated in generator mode by simply connecting a passive element at the coils terminals (a resistance), there is no control over the resulting current phase as would normally be the case for a motor. Finally, the same measures are carried out again, but with a change in the electrical characteristics of the coils. This is achieved by connecting external inductors, characterized by an inductance L_{add} and a resistance R_{add}, on each coil (Figure 16c). This means that the currents flowing through the coils are equal to, for the first case:

$$\bar{I}_{1,1} = -\bar{I}_{1,2} = \frac{\bar{E}_d}{R + j\omega L_c}. \quad (18)$$

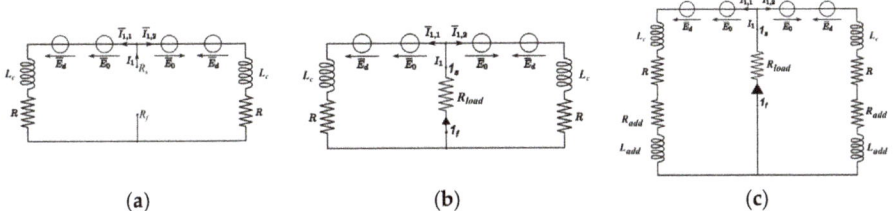

Figure 16. Electrical circuit for the experimental measurements with (a) the motor connections in open circuit, (b) a resistive charge at motor connections, and (c) with additional inductors and a resistive charge at the motor connections.

For the second case:

$$\bar{I}_{1,1} = \frac{\bar{E}_0}{R + 2R_{load} + j\omega L_c} + \frac{\bar{E}_d}{R + j\omega L_c}, \quad (19)$$

$$\bar{I}_{1,2} = \frac{\bar{E}_0}{R + 2R_{load} + j\omega L_c} - \frac{\bar{E}_d}{R + j\omega L_c}, \quad (20)$$

$$\bar{I}_1 = \frac{2\bar{E}_0}{R + 2R_{load} + j\omega L_c}. \quad (21)$$

And for the third case:

$$\bar{I}_{1,1} = \frac{\bar{E}_0}{R + R_{add} + 2R_{load} + j\omega(L_c + L_{add})} + \frac{\bar{E}_d}{R + R_{add} + j\omega(L_c + L_{add})}, \quad (22)$$

$$\bar{I}_{1,2} = \frac{\bar{E}_0}{R + R_{add} + 2R_{load} + j\omega(L_c + L_{add})} - \frac{\bar{E}_d}{R + R_{add} + j\omega(L_c + L_{add})}, \quad (23)$$

$$\bar{I}_1 = \frac{2\bar{E}_0}{R + R_{add} + 2R_{load} + j\omega(L_c + L_{add})}. \quad (24)$$

6.1. Bearing Forces

Referring to Equations (13) and (14), the electrodynamic bearing forces are proportional to the center shift. This is indeed observed in Figure 17, for the centering and perpendicular forces, as a function of the rotor displacement along the x-axis, (y-axis centered) for a spin speed of 6000 rpm. Both measures when the generator is left in open circuit, and when it is connected to a $R_{load} = 200\ \Omega$ charge, are presented in Figure 17.

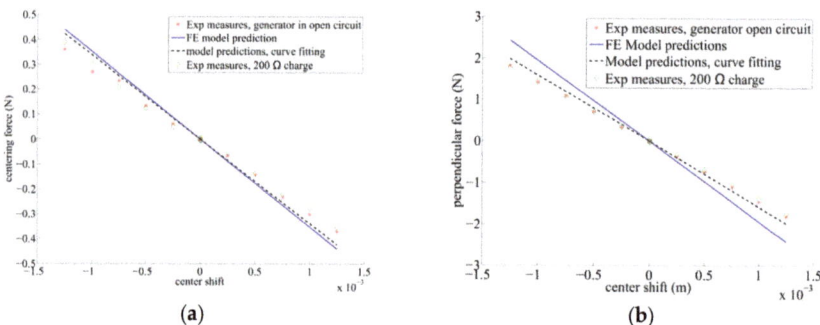

Figure 17. Bearing forces as a function of the rotor displacement (x-axis) for a spin speed of 6000 rpm. Measures when the generator is left in open circuit, and when connected to a 200 Ω charge, and comparison with the model predictions; (**a**) restoring force and (**b**) destabilizing force.

It can be observed in Figure 17 that the bearing forces are not influenced by generator behavior at the spin speed of 6000 rpm.

The comparison with the model predictions when its parameters are identified either by FE simulations or by curve fitting, is also shown, and there is a good correspondence. The values of the model parameters for both cases are given in Table 3, and are compared to the experimental measurement of the coil impedance.

Table 3. EMF model parameter identification.

Parameter	FE Value	Exp. Measure	Curve Fitting
$R(\Omega)$	44.11	42.08	42.1
$L_c(H)$	0.0127	0.0162	0.0143

The stiffness reached by the centering force at 6000 rpm can be deduced by the measures illustrated in Figure 17, and it is equal to 297.4 N/m, which is very low. However, the spin speed at which these experimental measures have been accomplished is relatively low for an electrodynamic bearing, as the electrodynamic stiffness developed by electrodynamic bearings increases with the spin speed. It can also be observed in Figure 17 that at the spin speed of 6000 rpm, the destabilizing parasitic force perpendicular to the center shift is more important than the restoring force; 1.8 N versus 0.36 N. This is again due to the fact that, at 6000 rpm, the rotor is not spinning fast enough. The spin speed can be compared to the electrical pole of the coils: the restoring force becomes predominant over the perpendicular force when the spin speed becomes higher than the inverse of the electric time constant of the system $> \omega_{RL}$, with $\omega_{RL} = \frac{R}{L_c} = 28 \times 10^3$ rpm [6]. The model predictions for this bearing show (Figure 18) that when extrapolating at higher spin speed, the perpendicular force decreases, and the restoring stiffness increases, as expected.

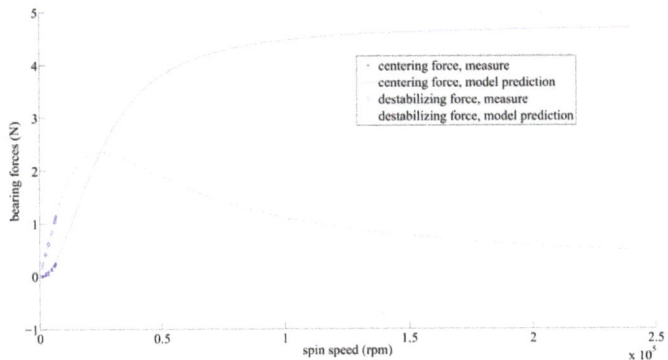

Figure 18. Bearing forces as a function of the spin speed for a center shift of 0.75 mm, experimental measures, and model predictions with model parameters identified by curve fitting.

Measures with Additional Inductors

In order to change the value of ω_{RL}, additional inductors have been connected in series with each coil of the prototype (see Figure 16c), the total value of these inductors is given in Table 4.

Table 4. Additional inductor values.

Case	$R_{add}(\Omega)$	$L_{add}(mH)$	$\omega_{RL}(rpm)$
One additional inductor	11.17	38.7	9600
Two additional inductors	17.01	79.1	6040

The bearing forces were measured with and without a resistive load at the motor terminals. First, it can be observed in Figure 19 that in this case, there is a good correspondence between the model predictions and the experimental measures. Moreover, the same observation as in Figure 17 on the coincidence of the measured force values with one additional inductor, with or without resistive charge at the motor terminals, can be obtained: the bearing forces are not influenced by generator behavior. In Figure 19, one can see that at the spin speed of 4800 rpm, the addition of external inductors allows the increase of the restoring component of the force while decreasing the destabilizing perpendicular bearing force. In Figure 20, at 6600 rpm, one can see that with the two additional inductors, the restoring component of the bearing forces becomes higher than the destabilizing one.

When observing the evolution of the bearing forces with the spin speed in Figure 21, one can see that the model predictions and measures are in agreement. With the external inductors, the spin speed ω_{RL} is modified and the spin speed above which the centering force becomes predominant is lower. This is confirmed by the experimental measures. However, one can also see that the asymptotic value that can be reached by the centering forces at high spin speed also decreases by adding the external inductors: the external inductors improve the centering force at low spin speed, which is good for our test bench as the spin speed is limited, but is harmful for high spin speed operation. In other words, the maximum stiffness the electrodynamic bearing is able to develop decreases when adding the external inductors.

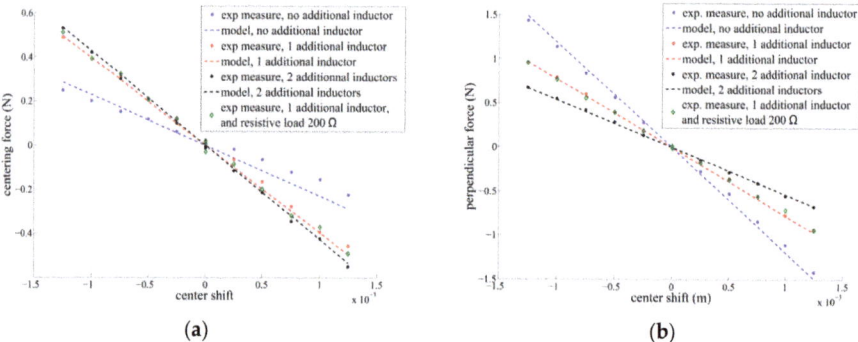

Figure 19. (a) Restoring force and (b) destabilizing bearing forces as a function of the rotor displacement along the x-axis while centered along the y-axis for a spin speed of 4800 rpm. Measures with and without additional inductors when the generator is left in open circuit and when connected to a 200 Ω charge, and comparison to the model when its parameters are identified by curve fitting.

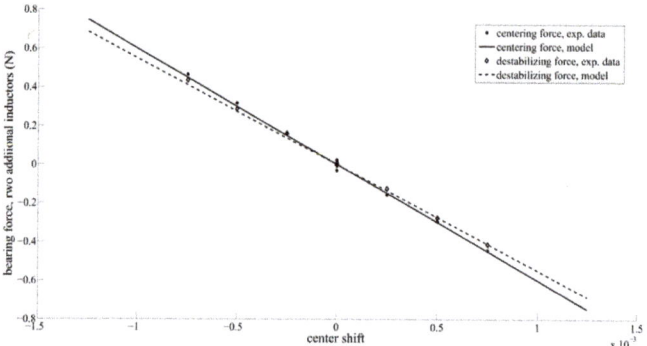

Figure 20. Bearing forces when adding two additional inductors in series with each coil as a function of the center shift for a spin speed of 6600 rpm.

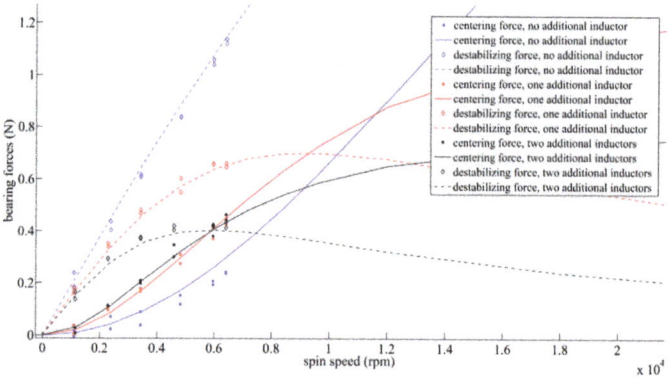

Figure 21. Bearing forces when adding additional inductors in series with each coil as a function of the spin speed for a center shift of 0.75 mm.

An additional word should be said on the well-known dynamic stability problems of centering electrodynamic bearings: the force perpendicular to the center shift will produce dynamic instability, and the system needs some damping in order to be dynamically stable when spinning. The dynamics of such a system are studied in detail through root loci in [14] for heteropolar electrodynamic bearings: they will depend on the R-L dynamics of the bearing winding and on the mechanical time constant of the system. Applying the same stability analysis as in [14] on the model presented in Equation (12) in the previous section, and knowing that the rotor mass is equal to 0.27 kg, Figure 22a,b show the root locus that has two stable roots, but one root which remains unstable for any spin speed. The amount of damping that should be added inside the system to achieve stable dynamic behavior can be studied theoretically through this root locus, as shown in Figure 22b: when adding progressive damping of [10 20 30] Ns/m, the spin speed above which the system is dynamically stable decreases to [10550 4460 1600] rad/s, respectively. The influence of the additional inductors can be observed in Figure 22c. They clearly have an influence on the spin speed above which the system is dynamically stable: with external damping of 20 and 30 Ns/m, the system is stable for any spin speed.

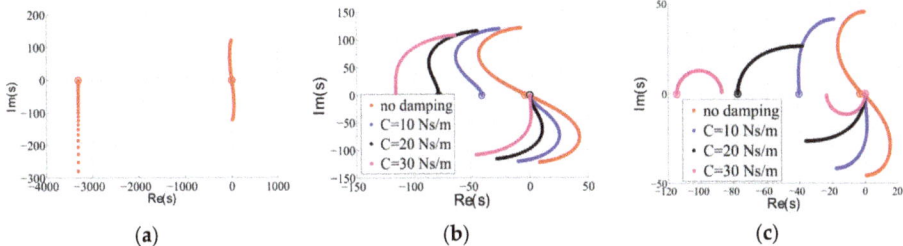

Figure 22. Theoretical root locus for the electrodynamic bearing prototype dynamic behavior. (**a**) All roots; (**b**) influence of external damping; and (**c**) theoretical root locus for the electrodynamic bearing prototype dynamic behavior when adding two external inductors and influence of external damping.

However, as all the measures presented in this article were performed in quasistatic conditions, these instabilities do not appear. By consequence, this test bench does not need to add this external damping, which remains a major issue for centering electrodynamic bearings.

6.2. Torque

The torque is first measured with the generator in open circuit and then with a resistive charge connected to its terminals. It can be observed in Figure 23a that when there is no charge on the motor terminal, torque is measured, which corresponds to the electrodynamic drag torque. This torque increases when the center shift increases. When the resistive charge is connected, an additional torque appears, corresponding to the motor torque. It can be observed that the motor torque is more important than the electrodynamic drag torque linked to the bearing function. This motor torque is proportional to the spin speed, as can be seen in Figure 23b, and remains constant while off-centering the rotor. In this figure, it can also be observed that when the rotor remains centered, the electrodynamic torque remains around zero, which confirms the null-flux principle: when the electrodynamic bearing is magnetically centered, the power dissipated inside the windings due to its bearing function is minimal, and no forces are generated.

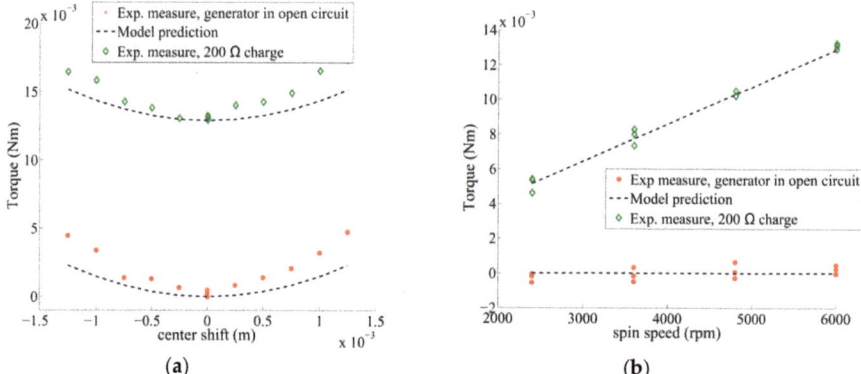

Figure 23. Measured and predicted torque without additional inductors (**a**) as a function of the rotor displacement along the x-axis, while centered along the y-axis for a spin speed of 6000 rpm and (**b**) as a function of the spin speed when the rotor is centered. Measures when the generator is left in open circuit and when connected to a 200 Ω charge.

The torque can be predicted by the state space model, as shown by Equation (15), but can also be predicted by performing a power balance: the power dissipated in the resistive load R_{load} corresponds to the generator output power, while the power dissipated by the current due to the center shift inside the coil resistances $R + R_{add}$ corresponds to the electrodynamic drag torque power. Finally, the power dissipated inside the coil resistances $R + R_{add}$ by the motor current corresponds to the motor Joule losses, and does not produce any torque. This leads to the relations:

$$\omega T_{ED} = 3 (R + R_{add}) \left(\frac{E_d}{R + R_{add} + j\omega(L_c + L_{add})} \right)^2, \quad (25)$$

$$\omega T_{motor} = 3 R_{load} \left(\frac{2E_0}{R + R_{add} + 2R_{load} + j\omega(L_c + L_{add})} \right)^2. \quad (26)$$

The electrodynamic torque model predictions with Equation (15) or (25) give the same results, and they match well with the measures in Figure 23a. The motor torque predicted by the power balance in Equation (26) also coincides with the experimental measures.

In Figure 24, it can be observed that the electrodynamic torque value is noticeably influenced by the additional inductors, while their influence is negligible on the motor torque. Finally, a global remark can be made on the measures shown: the output torque and the centering stiffness are weak. This can be explained by the fact that this electrodynamic heteropolar bearing was first solely intended to demonstrate the null-flux principle in heteropolar EDBs, and this prototype has not undergone any optimization of its performances in the sizing process.

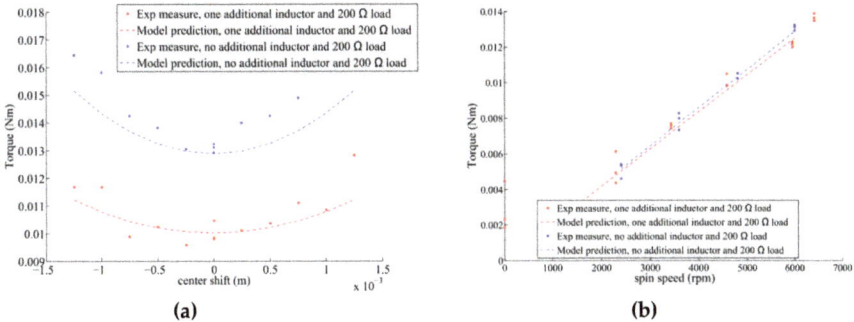

Figure 24. Measured and predicted torque, when the generator is connected to a 200 Ω charge (**a**) as a function of the rotor displacement along the x-axis while centered along the y-axis for a spin speed of 6000 rpm and (**b**) as a function of the spin speed when the rotor is centered. Measurements were conducted with and without external inductors.

6.3. Current

Finally, the current flowing through the coils has also been measured, and is shown in Figure 25 for one phase. It is compared to the current predicted by solving the electrical circuits of Figure 4 that are described in Equations (18–24).

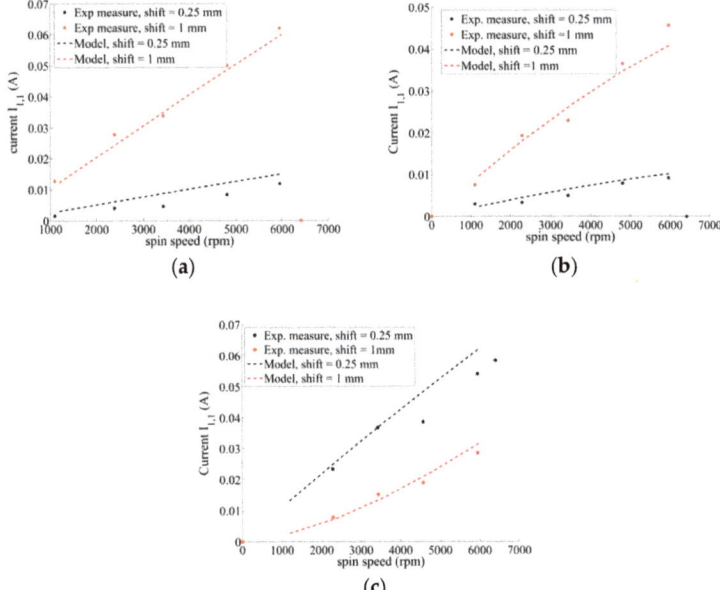

Figure 25. Measured and predicted current in first coil of first phase for center shifts of 0.25 and 1 mm in the case of the electrodynamic bearing prototype with (**a**) motor terminals in open circuit, (**b**) motor terminals in open circuit with one additional inductor on each coil, and (**c**) motor terminals with resistive load of 200 Ω and one additional inductor on each coil.

It can be noticed that in the two first cases, when the motor terminals are in open circuit, with or without additional inductors, in Figure 25a,b, the RMS value of the current in the first coil of phase 1 is more important for a center shift of 1 mm than 0.25 mm. Indeed, in these cases, there is no motor current, and the current inside the coils is only due to the bearing function. For the case with a resistive load at motor terminals and one additional inductor on each coil, in Figure 25c, one can notice that the current in the first coil is more important for a center shift of 0.25 mm than for a center shift of 1 mm. This is due to the fact that in this coil, the motor current and the bearing current flow in opposite directions. In contrast, in the second coil, the motor current and the bearing current add up, and the total current is more important for the higher center shift, as the model shows in Figure 26.

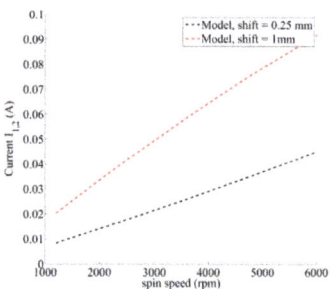

Figure 26. Predicted current in the second coil of the first phase for center shifts of 0.25 and 1 mm in the case of the electrodynamic bearing prototype with motor terminals with resistive load of 200 Ω and one additional inductor on each coil.

7. Conclusions

This paper shows experimental measures made on a heteropolar electrodynamic bearing to evaluate its ability to operate as a passively centered self-bearing permanent magnet rotor. First, the operating principle was briefly explained. Second, EMF measurements showed a motor EMF independent from the rotor center position and a bearing EMF proportional to the rotor displacement. Third, force measurements showed radial force measurements in agreement with electrodynamic centering bearing forces, presenting a restoring component and a component perpendicular to the center shift. It is shown that it is possible to shift the spin speed at which the electrodynamic bearing forces become predominantly centering forces towards lower spin speed by adding additional inductors in series on each coil of the prototype. It is also shown that the bearing forces are independent of the load connected to the motor terminals. Fourth, the torque measurements show an electrodynamic drag torque depending on the rotor position, and a motor torque proportional to the spin speed. The electrodynamic drag torque is influenced by the additional inductors, while the motor torque is almost not influenced by additional inductors. Finally, the current flowing in the coils was accurately predicted by the equivalent circuit models.

Despite the low produced torque and high stiffness, these experimental measures confirm the feasibility of the operating principle for a slotless winding without ferromagnetic yoke and a permanent magnet rotor. However, two major issues remain before realizing a completely passive self-bearing motor based on a radial electrodynamic bearing: the introduction of passive axial forces without impeding the radial stiffness developed by the radial electrodynamic bearing, and the introduction of sufficient damping in the system to counter the dynamic instabilities inherent to centering electrodynamic bearings.

Author Contributions: Writing—original draft, V.K.; Writing—review & editing, J.V.V. and B.D.

Funding: This research was funded by the university UCLouvain.

Conflicts of Interest: The authors declare no conflict of interest.

References

1. Severson, E.L.; Nilssen, R.; Undeland, T.; Mohan, N. Design of dual purpose no-voltage combined windings for bearingless motors. *IEEE Trans. Ind. Appl.* **2017**, *53*, 4368–4379. [CrossRef]
2. Xu, Z.; Lee, D.H.; Ahn, J.W. Comparative analysis of bearingless switched reluctance motors with decoupled suspending force control. *IEEE Trans. Ind. Appl.* **2014**, *51*, 733–743. [CrossRef]
3. Zhao, C.; Zhu, H. Design and analysis of a novel bearingless flux-switching permanent magnet motor. *IEEE Trans. Ind. Electron.* **2017**, *64*, 6127–6136. [CrossRef]
4. Gruber, W.; Rothböck, M.; Schöb, R.T. Design of a novel homopolar bearingless slice motor with reluctance rotor. *IEEE Trans. Ind. Appl.* **2014**, *51*, 1456–1464. [CrossRef]
5. Mitterhofer, H.; Mrak, B.; Gruber, W. Comparison of high-speed bearingless drive topologies with combined windings. *IEEE Trans. Ind. Appl.* **2014**, *51*, 2116–2122. [CrossRef]
6. Chiba, A.; Fukao, T.; Rahman, M.A. Vibration suppression of a flexible shaft with a simplified bearingless induction motor drive. *IEEE Trans. Ind. Appl.* **2008**, *44*, 745–752. [CrossRef]
7. Laiho, A.; Sinervo, A.; Orivuori, J.; Tammi, K.; Arkkio, A.; Zenger, K. Attenuation of harmonic rotor vibration in a cage rotor induction machine by a self-bearing force actuator. *IEEE Trans. Magn.* **2009**, *45*, 5388–5398. [CrossRef]
8. Xin, C.; Zhiquan, D.; Xiaolin, W. Vibration reduction with radial force windings short-circuited in bearingless switched reluctance motors. In Proceedings of the 2009 4th IEEE Conference on Industrial Electronics and Applications, Xi'an, China, 25–27 May 2009; pp. 2630–2634.
9. Kluyskens, V.; Dehez, B. Dynamical electromechanical model for magnetic bearings subject to eddy currents. *IEEE Trans. Magn.* **2012**, *49*, 1444–1452. [CrossRef]
10. Detoni, J.G. Progress on electrodynamic passive magnetic bearings for rotor levitation. *Proc. Inst. Mech. Eng. Part C J. Mech. Eng. Sci.* **2014**, *228*, 1829–1844. [CrossRef]
11. Amati, N.; de Lépine, X.; Tonoli, A. Modeling of Electrodynamic Bearings. *J. Vib. Acoust.* **2008**, *130*, 61007. [CrossRef]
12. Filatov, A.V.; Maslen, E.H. Passive magnetic bearing for flywheel energy storage systems. *IEEE Trans. Magn.* **2001**, *37*, 3913–3924. [CrossRef]
13. Dumont, C.; Kluyskens, V.; Dehez, B. Null-flux radial electrodynamic bearing. *IEEE Trans. Magn.* **2014**, *50*, 1–12. [CrossRef]
14. Dumont, C.; Kluyskens, V.; Dehez, B. Linear state-space representation of heteropolar electrodynamic bearings with radial magnetic field. *IEEE Trans. Magn.* **2015**, *52*, 1–9. [CrossRef]
15. Kluyskens, V.; Dumont, C.; Dehez, B. Description of an electrodynamic self-bearing permanent magnet machine. *IEEE Trans. Magn.* **2016**, *53*, 1–9. [CrossRef]
16. Kluyskens, V.; Dumont de Chassart, C.; Dehez, B. Design and experimental testing of a heteropolar electrodynamic bearing. In Proceedings of the 2th IEEE Conference on Advances in Magnetics, La Thuile, Italy, 4–7 February 2018.
17. Kluyskens, V.; van Verdeghem, J.; Dumont, C.; Dehez, B. Experimental Investigations on a Heteropolar Electrodynamic Bearing-self-bearing Motor. In Proceedings of the 16th International Symposium on Magnetic Bearings (ISMB), Beijing, China, 13–17 August 2018.

© 2019 by the authors. Licensee MDPI, Basel, Switzerland. This article is an open access article distributed under the terms and conditions of the Creative Commons Attribution (CC BY) license (http://creativecommons.org/licenses/by/4.0/).

Article

Space Vector Modulation Strategies for Self-Sensing Three-Phase Radial Active Magnetic Bearings [†]

Dominik Wimmer *, Markus Hutterer, Matthias Hofer and Manfred Schrödl

Institute of Energy Systems and Electrical Drives, Technische Universität Wien, Gußhausstr. 25, 1040 Vienna, Austria; markus.hutterer@tuwien.ac.at (M.H.); matthias.hofer@tuwien.ac.at (M.H.); manfred.schroedl@tuwien.ac.at (M.S.)
* Correspondence: dominik.wimmer@tuwien.ac.at
† This paper is an extended version of our paper published in: Wimmer, D.; Hutterer, M.; Hofer, M.; Schrödl, M. Variable Space Vector Modulation for Self-Sensing Magnetic Bearings. In Proceedings of the 16th International Symposium on Magnetic Bearings (ISMB16), Beijing, China, 13–17 August 2018.

Received: 19 April 2019; Accepted: 10 May 2019; Published: 14 May 2019

Abstract: The focus of this study lies on the investigation of the space vector modulation of a self-sensing three-phase radial active magnetic bearing. The determination of the rotor position information is performed by a current slope-based inductance measurement of the actuator coils. Therefore, a special pulse width modulation sequence is applied to the actuator coils by a conventional three-phase inverter. The choice of the modulation type is not unique and provides degrees of freedom for different modulation patterns, which are described in this work. For a self-sensing operation of the bearing, certain constraints of the space vector modulation must be considered. The approach of a variable space vector modulation is investigated to ensure sufficient dynamic in the current control as well as the suitability for a self-sensing operation with an accurate rotor position acquisition. Therefore, different space vector modulation strategies are considered in theory as well as proven in experiments on a radial magnetic bearing prototype. Finally, the performance of the self-sensing space vector modulation method is verified by an external position measurement system.

Keywords: active magnetic bearing; self-sensing; radial three-phase bearing; space vector modulation

1. Introduction

Magnetic bearings are of great significance for the stabilization of levitating rotors. Due to the fact that it is impossible to stabilize all degrees of freedom of a rigid body by permanent magnets, a force of a different physical origin must be applied for rotor stabilization [1,2]. Therefore, the rotor can be stabilized by electromagnets, which is stated as active magnetic bearing (AMB). AMBs require a position feedback information of the rotor to allow a stable operation. In this work, a self-sensing method is used to obtain the rotor position instead of using separate position sensors. The operation of self-sensing AMBs has been a field of research for many years [3–5] and provides advantages concerning sensor failure, construction space and production costs of the AMB. The self-sensing position determination of this study is performed by a modulation-based current switching ripple evaluation of the actuator coils. Previous studies presented different approaches for extracting the rotor position information out of the current ripple, such as current slope measurements [6,7], current ripple demodulation [8] or by the usage of artificial neural networks [9]. In this proposal, the so-called INFORM (Indirect Flux Detection by Online Reactance Measurement) method is used to determine the rotor position. This method was originally designed for the rotor angle determination of a permanent magnet synchronous motor [10]. Concerning AMBs, the INFORM method is based on a current slope measurement, detecting the inductance change depending on the rotor's eccentricity. As previously described in [11], the implementation of the self-sensing operation was based on an injection of voltage pulses to the actuator coils. Therefore, the current controller stops in equidistant time steps and measurement pulses are applied to the coils. This approach has the drawback of a limited bandwidth of the position measurement and an interruption of the current controller. To avoid this circumstance, the required pulse pattern for position measurement is embedded in the pulse width modulation (PWM) sequence of the current controller. The use of an embedded pulse pattern, in particular the 3-Active pattern [12], is the point of origin for the space vector modulation in this work.

2. Self-Sensing Bearing Setup

The bearing setup consists of two radial homopolar six pole AMBs with a common shaft as illustrated in Figure 1. The bias flux of the bearing is realized by the use of permanent magnets (PM). An axial displacement of the rotor is stabilized by a positive axial stiffness given by the bias flux and the geometry of the bearing (Figure 1b).

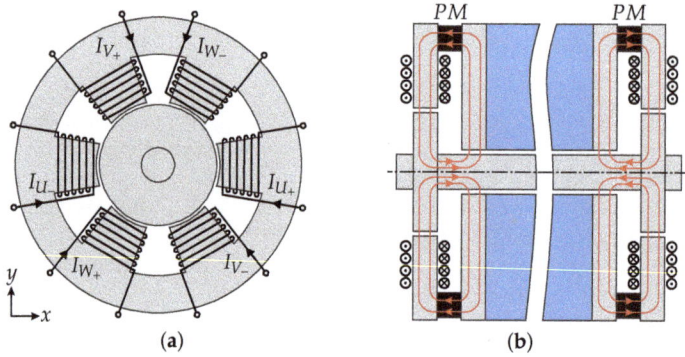

Figure 1. (a) Structural design of the six pole radial homopolar active magnetic bearing. (b) Cross section of the shaft: The bias flux (indicated by red lines) is generated by means of permanent magnets.

The stabilization of a radial rotor eccentricity is performed by the coils, which are driven in a differential configuration. The coils of the poles U_+, V_+, W_+ and the opposite coils are connected in wye-configuration. For achieving low system costs, a conventional three-phase inverter is aspired for the control of the AMB [13,14]. Two opposite coils are corresponding to one phase, which enables the use of a three-phase inverter. Figure 2 shows a differential transformer at the connection point of the positive and negative phase of the bearing.

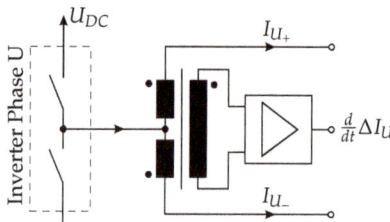

Figure 2. Differential current slope measurement by means of a differential open loop transformer.

By the usage of an open loop transformer in each phase, it is possible to connect the AMB to a three-phase inverter. The transformer can be realized as a separate unit or also be integrated in the printed circuit board (PCB) of the power electronics [15]. The transformer output provides the differential current slope signal, which is required for the self-sensing operation of the bearing. Figure 3a shows the current ripple caused by the PWM switching pattern. Although the stator and the rotor components were built from laminated iron sheets, it can be seen that the current slope is distorted by eddy currents. To overcome this problem, the current slope measurement is performed as a differential evaluation of two opposite coils to suppress the influence of eddy currents. The differential current slope evaluation is not limited to the homopolar design and can be also applied to heteropolar bearings [16]. Another approach would be a model-based consideration of the eddy currents as shown in [17]. Figure 3b shows the output voltage of the open loop transformer corresponding to the differential current $\Delta I_U = I_{U_+} - I_{U_-}$ of Figure 3a. After the settling time, the output voltage is proportional to the differential current slope $\frac{d}{dt}\Delta I_U$.

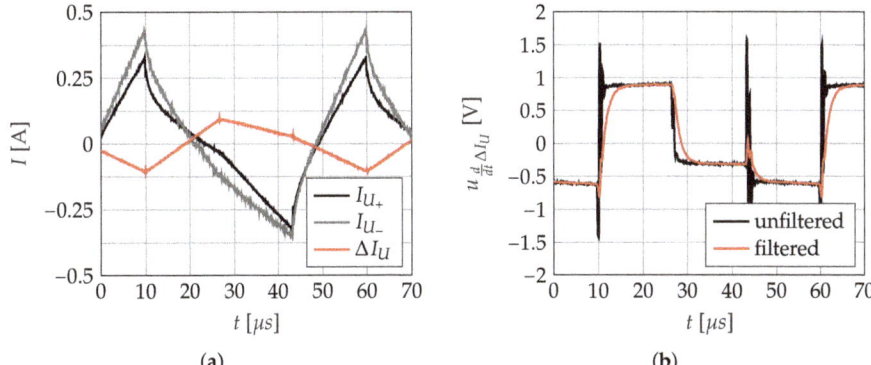

Figure 3. (a) Current ripple of the actuator coils. The influence of eddy currents is mostly suppressed in the differential current signal ΔI_U (f_{PWM} = 20 kHz, U_{DC} = 60 V). (b) Output signal of the open loop transformer corresponding to (a). The output voltage is proportional to the differential current slope.

Thus, the rotor position can be obtained by the approximation

$$u(t) = L(x,y)\frac{d}{dt}\Delta I(t) \rightarrow L(x,y) = u(t)\left(\frac{d}{dt}\Delta I(t)\right)^{-1} \quad (1)$$

with the coil voltage $u(t)$ and the position-depending inductance $L(x,y)$, which is described in [12].

3. Problem Formulation

Measurements on a prototype of a self-sensing radial AMB (Figure 4) have shown significant power losses in the bearing in the 3-Active PWM mode. Consequently, the power losses lead to a temperature rise in the bearing, which is undesirable in many applications.

Figure 4. Prototype of a self-sensing radial homopolar active magnetic bearing with six poles. The bias flux of the bearing is realized by the use of permanent magnets.

Beside the control current, the current ripple of the coils causes remarkable power losses in the AMB. Hence, major power losses in the prototype occur in the flux leading paths by means of iron losses in the laminated iron sheets. One possible solution for the reduction of the eddy current losses is the use of soft magnetic composites (SMC), which is described in [18]. However, SMC materials have drawbacks like a smaller permeability and mechanical limitations [19]. This work follows an approach, which is independent of the used material. Figure 5a shows the 3-Active SVM pattern with the corresponding power losses (Figure 5b) of the prototype as a function of the DC-link voltage U_{DC}.

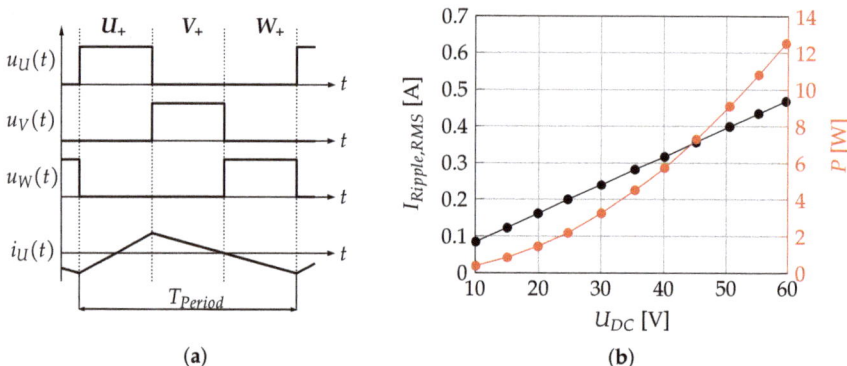

Figure 5. (a) 3-Active space vector modulation: Each PWM period contains voltage space vectors from each phase (U_+, V_+, W_+). (b) Power losses of the bearing due to the current ripple as a function of the DC-link voltage (3-Active SVM, f_{PWM} = 20 kHz).

For a fixed switching frequency, it is obvious to decrease U_{DC} for achieving small power losses. On the one hand, a high value of U_{DC} causes a high current ripple (Figure 5b), which provides a high magnitude of the differential current slope information. On the other hand, the iron losses are increased and therefore, undesirable power losses occur in the flux leading paths. To keep the power losses in the bearing small, a low level of U_{DC} is aspired. This circumstance gives the motivation to enhance the self-sensing position measurement for dealing with small current slopes. Furthermore, it must be considered that the dynamic of the current controller depends on U_{DC}. The focus of this work lies on the investigation of different space vector modulations to ensure a sufficient current dynamic as well as a high quality of the position measurement. The design of SVM contains degrees of freedom, which can be used for the development of specific pulse patterns with especially high current dynamics or high quality of the position measurement. Taking this one step further, it is possible to use different kinds of SVM during the operation of the AMB. This leads to the variable SVM and enables the combination of the properties of different modulations.

4. Space Vector Modulation

Figure 6 shows the symmetrical arrangement of the magnetic poles of the bearing in the xy-plane. The spatially distributed arrangement of the poles enables the definition of six fundamental voltage space vectors, which are aligned with the magnetic poles of the bearing. The voltage space vectors can be obtained by a conventional power inverter with three half bridges by the switching states shown in Table 1 [20].

Table 1. Voltage space vector definition of a three-phase inverter.

Phase	Switch	Voltage Space Vector							
		U+	U−	V+	V−	W+	W−	Z+	Z−
U	HS	1	0	0	1	0	1	1	0
U	LS	0	1	1	0	1	0	0	1
V	HS	0	1	1	0	0	1	1	0
V	LS	1	0	1	1	1	0	0	1
W	HS	0	1	0	1	1	0	1	0
W	LS	1	0	1	0	0	1	0	1

HS = High-Side switch, LS = Low-Side switch; 1 = closed, 0 = open.

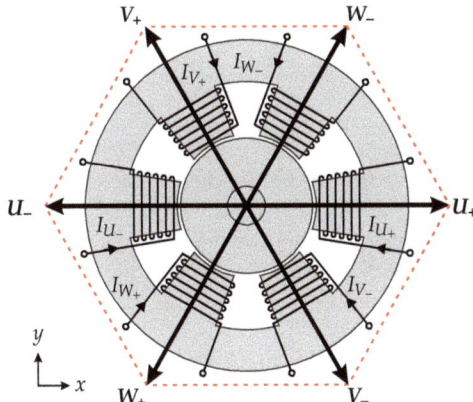

Figure 6. Cross section of the six pole homopolar radial AMB. The fundamental voltage space vectors U_+, U_-, V_+, V_-, W_+, W_- are aligned with the poles.

The fundamental space vectors shape a hexagon, which defines the possible modulation range of the voltage space vectors. Each point in this hexagon can be reached by a linear combination of six fundamental space vectors (U_+, U_-, V_+, V_-, W_+, W_-) and two zero space vectors (Z_+, Z_-). The zero space vector occurs if either all high-side or low-side switches of the three-phase inverter are closed. The calculation between the reference coordinate system (x, y) and the three phase system (U, V, W) is done by the Clarke-transformation [21]. The degree of freedom in the composition of the linear combination of the space vectors is restricted by the following design rules for self-sensing operation.

4.1. Design Rules for SVM

The design criteria of the SVM achieve a good quality of the position measurement and a high current controller bandwidth. Therefore, the design of the SVM underlies certain restrictions to allow a self-sensing operation of the magnetic bearing:

- **INFORM method:** Theoretically, the current ripple caused by a single voltage pulse contains the whole information of the rotor position. As asymmetries appear in the real system (caused by mechanical, electrical or magnetic deviations), it is beneficial to use the current slope information from independent voltage pulses. Furthermore, the voltage pulses must have a minimal pulse duration t_{INF}, which is given by the settling time of the current slope measurement path (Figure 3b). The duration is defined by the decay of the eddy currents and the settling of the analog filter, which causes a distortion of the differential current slope signal.

- **Modulation amplitude:** The modulation amplitude defines the maximum length of a desired voltage space vector. High modulation amplitudes of the desired voltage space vector allow a high dynamic of the current control. For a symmetrical (angle independent) operation of the current controller, it is beneficial to limit the modulation amplitude to the in-circle of the possible modulation area. Theoretically, the symmetrical modulation amplitude can achieve a value of $R_{max} = \sqrt{3}/2 \approx 0.866$ for the given AMB system.

- **Inverter:** Short pulse lengths could cause problems in semiconductor switches. Hence, the specified recovery time of the switches must be considered in the PWM pattern [22]. Furthermore, a proper operation of a potential charge pump of the gate driver must be ensured. Therefore, the pulse pattern has to provide at least one switching action in each phase.

The design procedure of the SVM is based on a representation of the desired voltage space vector P, which is specified by the superior current controller. Hence, the desired space vector P is given by means of all eight voltage space vectors.

$$P = c_{U_+}U_+ + c_{U_-}U_- + c_{V_+}V_+ + c_{V_-}V_- + c_{W_+}W_+ + c_{W_-}W_- + c_{Z_+}Z_+ + c_{Z_-}Z_- \quad (2)$$

Therefore, the coefficients c_{U_+}, c_{U_-}, c_{V_+}, c_{V_-}, c_{W_+}, c_{W_-}, c_{Z_+}, c_{Z_-} define the length of the corresponding space vector. The length of the respective voltage space vector is related to the duration of the voltage pulse in the PWM pattern. The following introduced variants of SVM differ substantially in the number of the active space vectors within one PWM period. In this context, "active" refers to the fundamental space vectors with an amplitude unlike zero. In the following considerations the minimum pulse duration t_{INF} is normalized to the PWM period.

4.2. 6-Active SVM

The 6-Active SVM uses all fundamental space vectors for the formation of P. Hence, P is built by a linear combination of the adjoining fundamental voltage space vectors. The remaining time of the PWM period is equally distributed to all fundamental voltage space vectors to build a combined zero space vector. The maximum modulation amplitude is obtained, if the length of one active space vector drops under t_{INF} and violates the timing requirements of a differential current slope measurement. Although the admissible modulation area is defined by the solid hexagon in Figure 7, the intended modulation area is limited by the minimum and maximum symmetrical modulation amplitude (R_{min}, R_{max}).

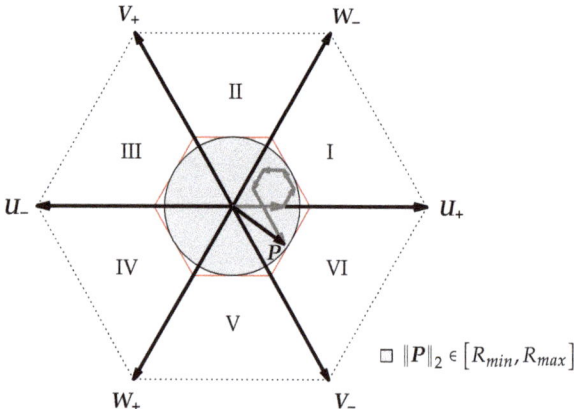

Figure 7. 6-Active SVM: The space vector P is formed by six fundamental space vectors ($t_{INF} = 0.1$).

The 6-Active SVM allows six independent current slope measurements within a PWM period. For this reason, the 6-Active SVM is well suited for the self-sensing method. However, this kind of modulation has the drawback of a very limited modulation amplitude R_{max}. The impact of the small value of R_{max} is caused by the minimal pulse width t_{INF} by means of Equations (3) and (4).

$$R_{max} = \frac{\sqrt{3}}{2}(1 - 6\,t_{INF}), \quad t_{INF} < \frac{1}{6} \tag{3}$$

$$R_{min} = 0 \tag{4}$$

For achieving a higher value of R_{max}, the number of space vectors is reduced in the following considerations.

4.3. 3-Active Low Dynamic Range SVM

In contrast to the 6-Active SVM, the 3-Active Low Dynamic Range (LDR) SVM uses either three positive (U_+, V_+, W_+) or three negative fundamental space vectors (U_-, V_-, W_-).

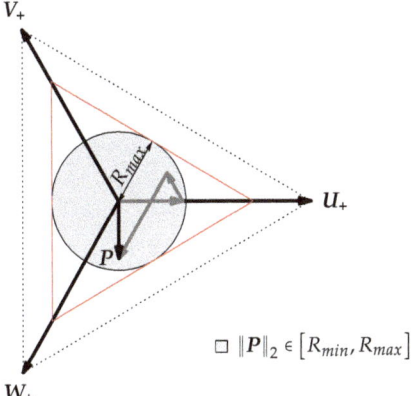

Figure 8. 3-Active LDR SVM: The space vector P is formed by the positive fundamental space vectors (U_+, V_+, W_+). ($t_{INF} = 0.1$).

Figure 8 shows a 3-Active LDR SVM with a linear combination of U_+, V_+, W_+. The corresponding coefficients from Equation (2) can be calculated by the inner product of P and the respective fundamental voltage space vector (Equations (5)–(7)).

$$c_{U_+} = \frac{1}{3} + \frac{2}{3} P \cdot U_+ \tag{5}$$

$$c_{V_+} = \frac{1}{3} + \frac{2}{3} P \cdot V_+ \tag{6}$$

$$c_{W_+} = \frac{1}{3} + \frac{2}{3} P \cdot W_+ \tag{7}$$

By using only three fundamental space vectors, the maximum modulation amplitude is limited to 0.5 and causes a smaller drop of R_{max}

$$R_{max} = \frac{1}{2}(1 - 3\,t_{INF}), \quad t_{INF} < \frac{1}{3} \tag{8}$$

$$R_{min} = 0 \tag{9}$$

by an increase of t_{INF} than the 6-Active SVM.

4.4. 4-Active SVM

The 4-Active SVM is based on the 3-Active LDR SVM, but enhances the modulation amplitude by means of an additional fundamental voltage space vector, which is located next to the desired space vector P like shown in Figure 9. For an effective implementation, the desired space vector P is built by a linear combination of the adjoining fundamental voltage space vectors. The remaining time of the PWM period is distributed equally to the positive (U_+, V_+, W_+) or negative (U_-, V_-, W_-) fundamental voltage space vectors.

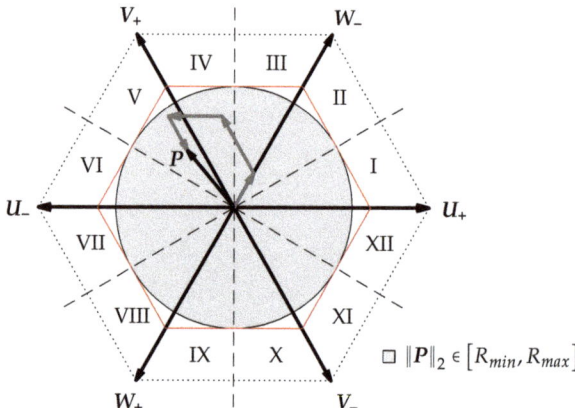

Figure 9. 4-Active SVM: The space vector P is formed by the negative (U_-, V_-, W_-) fundamental space vectors and V_+ for an enhancement of the modulation amplitude ($t_{INF} = 0.1$).

The limits of the symmetrical modulation amplitude

$$R_{max} = \frac{\sqrt{3}}{2}(1 - 3\, t_{INF}), \quad t_{INF} < \frac{1}{5} \qquad (10)$$

$$R_{min} = 0 \qquad (11)$$

are obtained if the length of one vector falls below t_{INF}. In contrast to the 3-Active LDR SVM, the maximum modulation amplitude is enhanced by a factor of $\sqrt{3}$ (Equation (10)).

4.5. 3-Active High Dynamic Range SVM

The 3-Active High Dynamic Range (HDR) SVM is designed for a maximum modulation amplitude using three fundamental space vectors and one of the zero space vectors (Z_+, Z_-). The desired space vector P is mainly formed by the two adjoining fundamental space vectors (V_+, W_- in Figure 10).

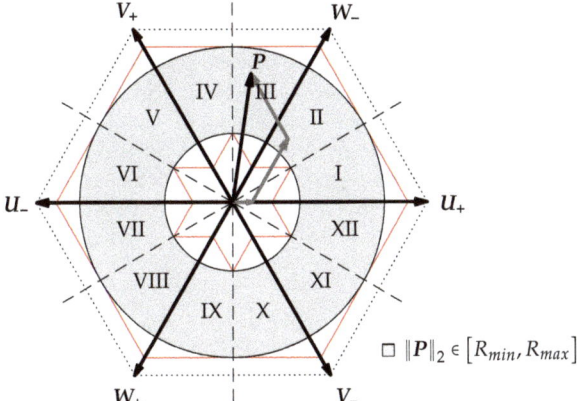

Figure 10. 3-Active HDR SVM: The desired space vector P is built with three fundamental space vectors (involving a space vector from each phase) and one zero space vector (t_{INF} = 0.1).

In order to get a current slope information by a voltage space vector from all space axis, the third fundamental space vector (U_+ in Figure 10) is added with the minimum length t_{INF}. A zero space vector fills the remaining time of the pulse pattern and ensures the required switching action in each phase. Equations (12) and (13) show the possible modulation range

$$R_{max} = \frac{\sqrt{3}}{2}(1 - t_{INF}), \; t_{INF} < \frac{1}{5} \tag{12}$$

$$R_{min} = 2\sqrt{3}\, t_{INF} \tag{13}$$

for a symmetrical operation.

4.6. Combination of the SVMs

The design of the SVMs shows, that each SVM has individual characteristics concerning the modulation amplitude and the usability for self-sensing operation. For an optimal operation of the bearing, it is possible to combine the properties of different SVMs. Figure 11 shows a comparison of the symmetrical modulation amplitudes as a function of t_{INF}.

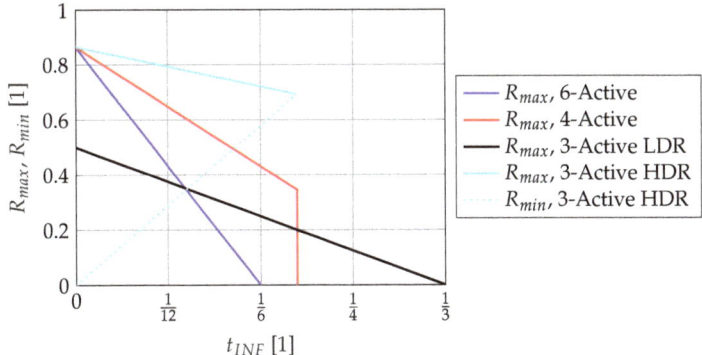

Figure 11. Comparison of the symmetrical modulation amplitude.

The 3-Active HDR has the property of a lower boundary R_{min} of the modulation amplitude. It is obvious, that a low value of t_{INF} leads to a high maximum modulation amplitude. The 3-Active LDR can operate up to $t_{INF} < 1/3$, which can be advantageous for applications with a high switching frequency. Concerning the performance of the self-sensing operation, each fundamental space vector gives additional information for the rotor position.

Figure 12 shows a combination of multiple SVMs for $t_{INF} = 0.1$. The choice of the SVM is determined in a way, that the desired space vector P is built by the SVM, which provides the highest number of active space vectors within a PWM period. To avoid nonessential switching between different SVMs, a hysteresis can be defined by means of an overlapping modulation area.

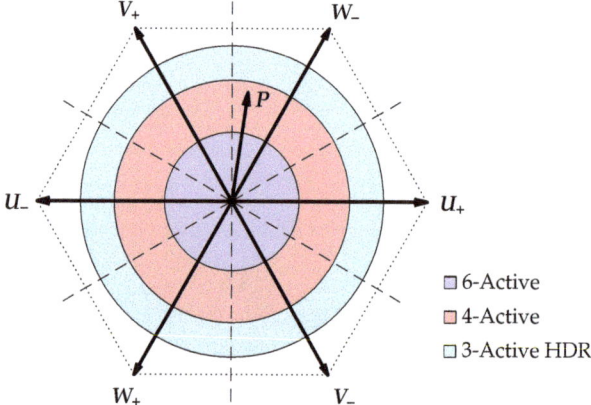

Figure 12. Combination of different SVMs during operation for achieving a high modulation area and a maximum performance of the self-sensing position measurement ($t_{INF} = 0.1$).

4.7. SVM Switchover

It is possible to make a distinction between two scenarios of SVM switchover. The first scenario is a sector switchover within a SVM. In the 4-Active and 3-Active HDR SVM the fundamental voltage space vectors used depend on the sector of P. Thus, the fundamental space vector changes if P changes to a new sector of the modulation area. Therefore, a switchover of the sector causes a change in the current ripple profile. Although the steady state of the mean value of the current is not affected by this effect, the phase currents obtain a transient error by a sector switchover. The amplitude of the current error is in the range of the amplitude of the current ripple. Figure 13a shows a simulation of a sector switchover for the 4-Active SVM for a space vector with $\|P\|_2 = 0$ and $\varphi = \pi/6$. It can be seen that the mean values of the phase currents differ after the sector switchover, which result in a transient deviation that decays over time.

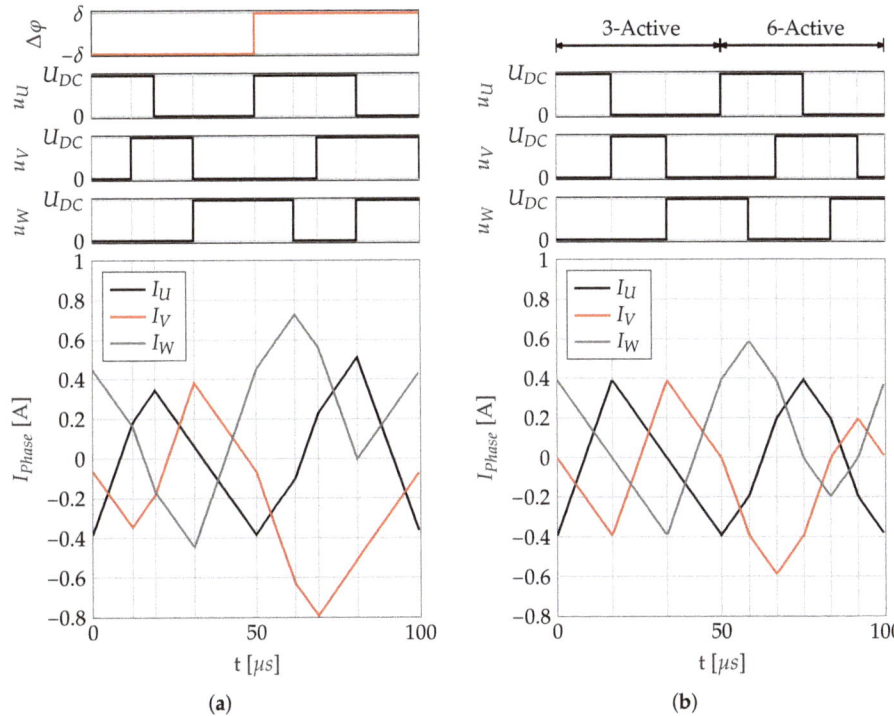

Figure 13. (a) Transient simulation of a sector switchover within the 4-Active SVM. The voltage space vector has an amplitude of zero but changes in the angle $\varphi = \pi/6 + \Delta\varphi$ to force a sector switchover. (b) Transient simulation of a switchover between the 3-Active and 6-Active SVM. Both SVMs represent a voltage space vector with zero amplitude (f_{PWM} = 20 kHz, t_{INF} = 0.14, U_{DC} = 48 V, δ = 0.001 rad).

The second scenario is given by a switchover between different SVMs. Figure 13b shows a switchover between the 3-Active LDR and the 6-Active SVM for the voltage space vector $\|\boldsymbol{P}\|_2 = 0$ and $\varphi = 0$. Although both SVM represent a zero space vector, a drift of the mean values of the currents can be observed after the switchover. This effect is caused by the different current waveforms of the SVMs. In many applications, the current ripple is significantly smaller than the control current and the drift due to sector switchover is negligible. However, if the application requires a precise current control beneath the amplitude of the current ripple, a compensation strategy is required. One conceivable solution would be the introduction of a modified PWM cycle to suppress the current drift after a SVM switchover.

5. Measurements

The following measurements compare the behavior of the designed SVMs, regarding power losses, dynamic of the current controller as well as the quality of the self-sensing position measurement. The measurements were performed on the prototype presented in Figure 4 applying the test setup of Figure 14. External eddy current-based position sensors were used as a reference for position measurements.

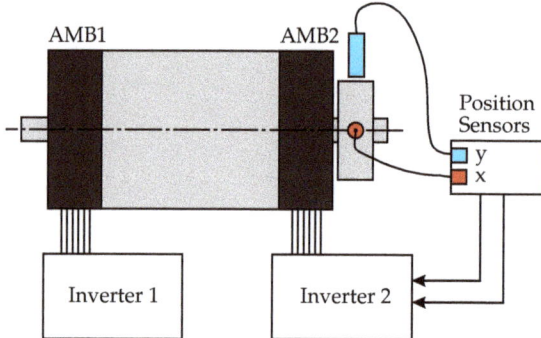

Figure 14. Symbolic arrangement of the test setup with external position sensors according to Figure 4.

The test setup with two homopolar radial magnetic bearings was controlled by independent three-phase inverters. Basically, a decoupled control of the rotor gives many degrees of freedom for advanced control [23]. However, simple decentralized PIDT1 position controllers provided sufficient performance for the following considerations. Concerning position control, it was assumed that the force on the rotor is proportional to the phase currents. In general, the currents of opposite coils (e.g., I_{U_+}, I_{U_-}) are not equal due to different inductances of an eccentrically levitating rotor. Therefore, the implemented force control by means of a static current force characteristic [24] is an approximation, but it does not cause any restrictions for the subsequent measurements.

5.1. Power Losses of the SVM Variants

The initial aim was to decrease the power losses in the bearing by a reduction of the DC-link voltage. Figure 15 shows a comparison of the power losses in the bearing for different SVMs at 20 kHz switching frequency. There is no significant difference between the SVMs, which allows an almost power neutral switchover between different SVMs.

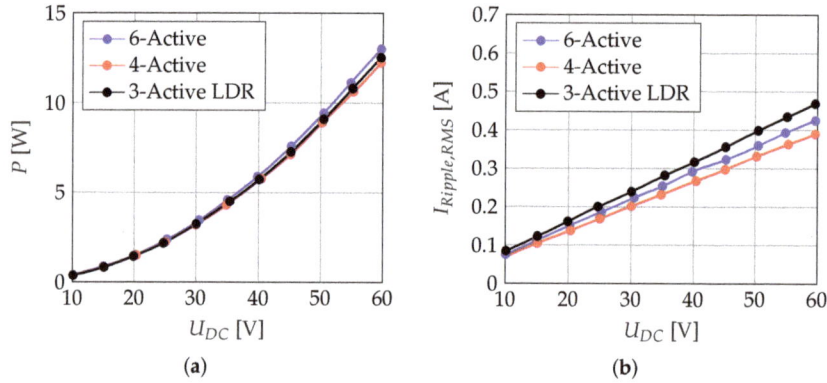

Figure 15. Comparison of the power losses in the bearing (**a**) and the current ripple of a phase (**b**) as a function of the DC-link voltage (f_{PWM} = 20 kHz, t_{inf} = 0.14).

5.2. Dynamic of the Current Controller

The maximum modulation amplitude R_{max} of the particular SVM has a significant impact on the dynamic of the current controller. Figure 16a shows a dynamic comparison of the current controller by means of the step response. Due to the fact that the 3-Active HDR SVM is not able to represent a modulation amplitude smaller than R_{min} (Figure 11), it is combined with the 4-Active SVM to obtain a steady state without oscillation.

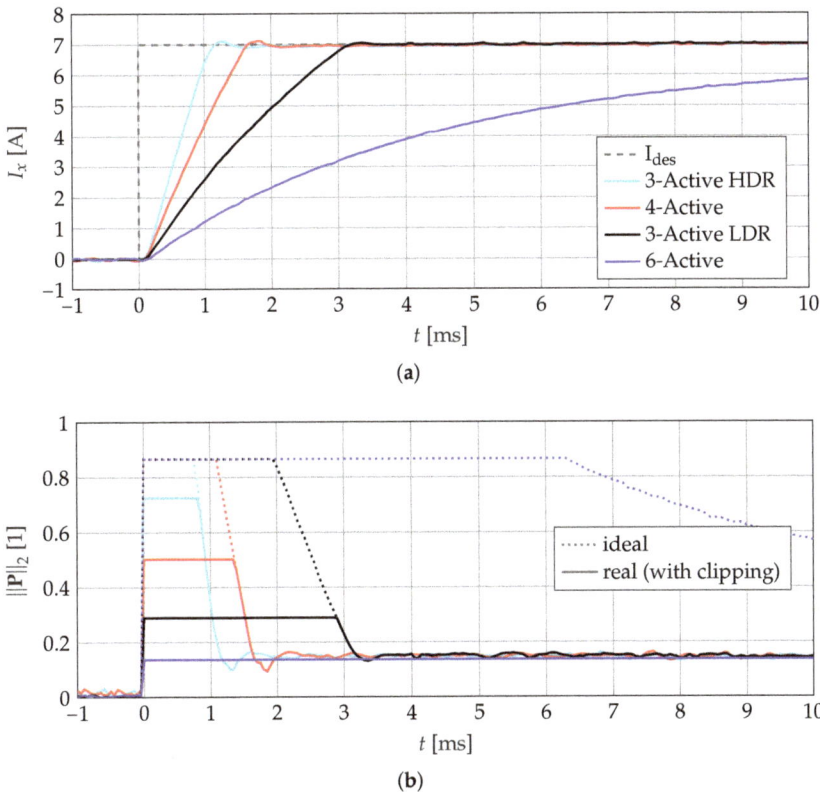

Figure 16. (a) Step response of the current controller for different SVMs. (b) Corresponding magnitude of the desired voltage space vector P (f_{PWM} = 20 kHz, U_{DC} = 12 V, t_{inf} = 0.14).

Figure 16b indicates the amplitude of the desired voltage space vector P, which corresponds to the step response. It can be seen, that the 3-Active HDR SVM has the highest modulation amplitude. Thus, the controller is only saturated for a short period of time at R_{max}. Concerning Figure 16, the 6-Active SVM is not able to inject the desired current of 7 A to the coil. The reason is the coil resistance and the connector cable. In this case, the 6-Active modulation can only be used for small currents and shows the use case for switching to a SVM with a higher modulation amplitude.

5.3. Noise of the Self-Sensing Method

The noise of the self-sensing method is of great significance for a precise control of the bearing. Figure 17a shows a noise comparison of the different SVMs with U_{DC} in the range of 20 V to 60 V. The 6-Active SVM has the lowest noise level, which is caused by the high number of voltage space vectors within a PWM period. Basically, the noise level increases proportional to $\propto U_{DC}^{-1}$. However, it is possible to shift the noise level in a certain range by an adaption of the analog circuit. Therefore, a low noise level can be achieved even at low values of U_{DC}. Thus, Figure 17b shows a similar noise characteristic like Figure 17a at the half DC-link voltage level (e.g., $\sigma_x < 0.15$ μm at 19 V) by means of an adaption of the analog circuit.

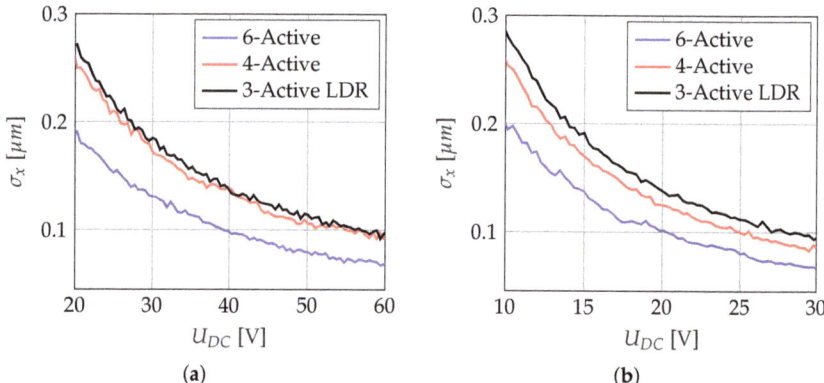

Figure 17. (a) Noise comparison of the self-sensing position measurement. (b) Reduction of the noise at low levels of U_{DC} by an adaption of the analog circuit (f_{PWM} = 20 kHz, rotor fixed in the center of the bearing, standard deviation σ_x over 3000 samples of the x-position at each setpoint of U_{DC}).

5.4. Linearity of the Self-Sensing Method

Beside the noise level, the self-sensing method must provide a high linearity for a proper operation of the bearing. The AMB prototype has an air gap of 800 µm and the auxiliary bearing allows a radial operational range of 400 µm. Figure 18 shows a linearity analysis of the 6-Active SVM with external position sensors for different rotor setpoints.

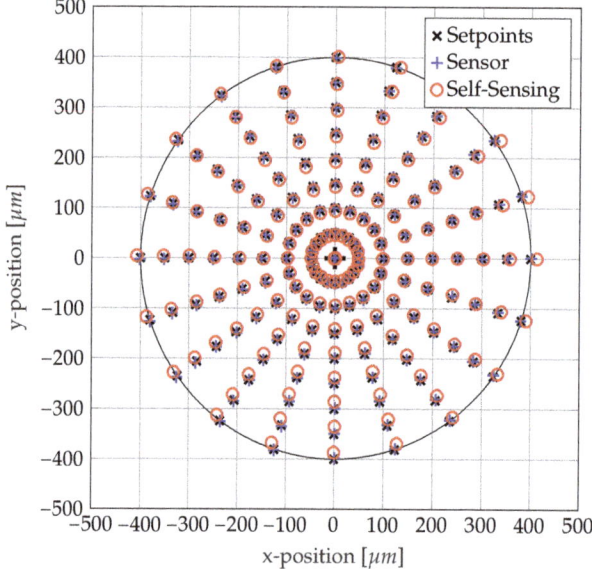

Figure 18. Linearity analysis with external position sensors for different setpoints of the rotor position (f_{PWM} = 20 kHz, 6-Active SVM).

The external sensors measure the rotor position on an aluminum disc, which is mounted close to the magnetic bearing. Due to an axial and angular shift of the sensors with regard to the AMB, the position information of the sensors is transformed in the self-sensing coordinate system. Figure 19 shows the self-sensing linearity error $e = \sqrt{e_x^2 + e_y^2}$ for different rotor positions for a SVM with 3 and 6 active voltage space vectors. It can be seen that the SVMs have a similar distribution of the linearity error with a maximum linearity error of about 16 µm.

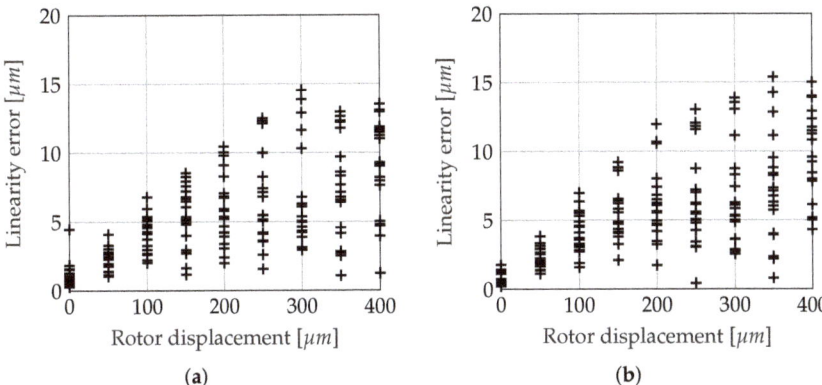

Figure 19. Analysis of the self-sensing linearity error $e = \sqrt{e_x^2 + e_y^2}$ as a function of the radial rotor displacement for different SVMs: (a) 3-Active LDR, (b) 6-Active.

The linearity of the self-sensing method is important for a proper control of the bearing, especially for robustness considerations of self-sensing magnetic bearings [25,26].

5.5. Small Signal Behavior

Finally, a comparison of the small signal behavior of the self-sensing method was performed. For this purpose, a disturbance is applied to the self-sensing levitating rotor and the sensor signal is shown for comparison (Figure 20). Both signals show a consistent behavior, which manifests a precise self-sensing operation.

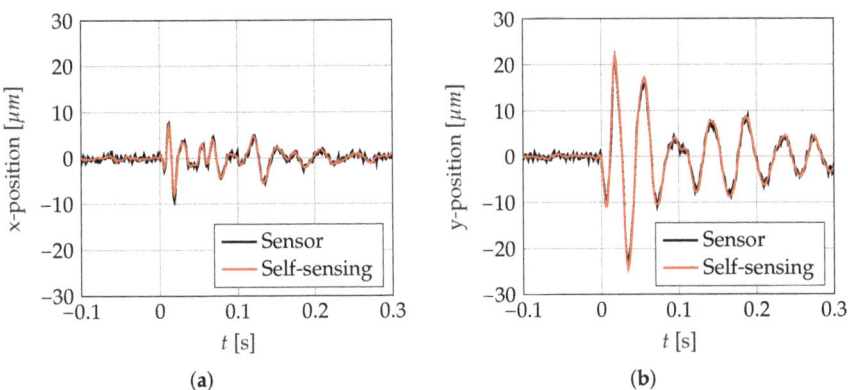

Figure 20. Comparison of the small signal behavior between an external sensor and the self-sensing method. A disturbance is applied at $t = 0$ s (a) x-position (b) y-position (f_{PWM} = 20 kHz, 6-Active SVM).

6. Results

The measurements on the prototype demonstrated the characteristics of different space vector modulations concerning self-sensing and current dynamic. The combination of different kinds of modulation allows a high modulation amplitude of voltage space vectors. This feature enables the possibility of a reduction of the DC-link voltage, while still having sufficient dynamic in the current control. Measurements on a prototype showed, that AMB losses due to the current ripple have an approximate square dependence of the DC-link voltage. Therefore, already a small reduction of the DC-link voltage can reduce power losses in the AMB. Furthermore, a measurement of the power losses revealed that the power losses caused by the switching ripple are nearly independent of the used SVM, which enables almost power neutral SVM switchover. The combination of different SVMs leads to a high dynamic of the current controller, which was proven by the step response of the system. An analysis of the switchover characteristics between different SVMs indicated, that a SVM switchover has an impact to the phase currents in the range of the ripple amplitude. This effect may lead to disturbances in the current control and could be suppressed by a compensation cycle in the SVM after a switchover. Regarding the self-sensing position measurement, the experimental results of the prototype showed a low noise level, which varied with the used SVM. A linearity analysis with external sensors showed an overall error of about 16 µm for different setpoints in the admissible rotor orbit.

7. Conclusions and Outlook

The focus of this study lied on an improvement of the space vector modulation of a three-phase self-sensing radial magnetic bearing. Different kinds of space vector modulations for self-sensing operation were presented in theory as well as proven in experiments on a prototype of a radial active magnet bearing. The results showed that the dynamic of the current controller and the quality of the self-sensing position control can be adjusted by the choice of the space vector modulation. The self-sensing method showed a high linearity and low noise of the rotor position for low control currents. In a next step, saturation effects of the flux leading paths will be considered regarding nonlinearities of the self-sensing operation to analyze the limitations of the self-sensing control. Further investigations will be performed at various rotational speeds to reveal potential speed-depending effects of the self-sensing control, especially with respect to a robust control of the system.

Author Contributions: Conceptualization, D.W., M.H. (Markus Hutterer), M.S.; methodology, D.W., M.H. (Markus Hutterer); software, D.W.; validation, D.W., M.H. (Markus Hutterer); formal analysis, D.W., M.H. (Markus Hutterer), M.H. (Matthias Hofer); investigation, D.W.; resources, M.S.; writing—original draft preparation, D.W.; writing—review and editing, M.H. (Markus Hutterer), M.H. (Matthias Hofer), M.S.; supervision, M.S.;

Funding: This research received no external funding.

Acknowledgments: This project was supported by Technische Universität Wien in the framework of the Open Access Publishing Program.

Conflicts of Interest: The authors declare no conflict of interest.

Abbreviations

The following abbreviations are used in this manuscript:

AMB	Active Magnetic Bearing
DC	Direct Current
HDR	High Dynamic Range
LDR	Low Dynamic Range
PCB	Printed Circuit Board
PM	Permanent Magnet

PWM Pulse Width Modulation
SMC Soft Magnetic Composite
SVM Space Vector Modulation

References

1. Schweitzer, G.; Maslen, E.H. *Magnetic Bearings: Theory, Design, and Application to Rotating Machinery*; Springer: Berlin, Germany; 2009; pp. 127–145, ISBN 978-3-642-00496-4
2. Earnshaw, S. On the Nature of Molecular Forces which Regulate the Constitution of the Lumiferous Ether. *Trans. Camb. Philos. Soc.* **1842**, *7*, 97-112.
3. Sivadasan K.K. Analysis of Self-Sensing Active Magnetic Bearings Working on Inductance Measurement Principle. *IEEE Trans. Magn.* **1996**, *32*, 329–334. [CrossRef]
4. Maslen, E. Self–sensing for active magnetic bearings: Overview and status. In Proceedings of the 10th International Symposium on Magnetic Bearings (ISMB10), Martigny, Switzerland, 21–23 August 2006.
5. Schammass, A.; Herzog, R.; Buhler, P.; Bleuler, H. New results for Self-Sensing Active Magnetic Bearings Using Modulation Approach. *IEEE Trans. Control. Syst. Technol.* **2005**, *13*, 509–516. [CrossRef]
6. Hofer, M.; Nenning, T.; Hutterer, M.; Schrödl, M. Current Slope Measurement Strategies for Sensorless Control of a Three Phase Radial Active Magnetic Bearing. In Proceedings of the 22nd International Conference on Magnetic Levitated Systems and Linear Drives (MAGLEV 2014), Rio de Janeiro, Brazil, 28 September–1 October 2014.
7. Wang, J.; Binder, A. Position estimation for self-sensing magnetic bearings based on the current slope due to the switching amplifier. *Eur. Power Electron. Drives (EPE)* **2016**, *26*, 125–141. [CrossRef]
8. Gruber, W.; Pichler, M.; Rothböck, M.; Amrhein, W. Self-Sensing Active Magnetic Bearing Using 2-Level PWM Current Ripple Demodulation. In Proceedings of the 7th International Conference on Sensing Technology (ICST), Wellington, New Zealand, 3–5 December 2013.
9. You, S.J.; Ahn, H.J. Position Estimation for Self-Sensing Magnetic Bearings using Artificial Neural Network. In Proceedings of the 16th International Symposium on Magnetic Bearings (ISMB16), Beijing, China, 13–17 August 2018.
10. Schrödl, M. Sensorless control of permanent-magnet synchronous machines at arbitrary operating points using a modified INFORM flux model. *Eur. Trans. Electr. Power ETEP* **1993**, *3*, 277–283. [CrossRef]
11. Hofer, M.; Hutterer, M.; Nenning, T.; Schrödl, M. Improved Sensorless Control of a Modular Three Phase Radial Active Magnetic Bearing. In Proceedings of the 14th International Symposium on Magnetic Bearings (ISMB14), Linz, Austria, 11–14 August 2014.
12. Nenning T.; Hofer M.; Hutterer M.; Schrödl M. Setup with two Self-Sensing Magnetic Bearings using Differential 3-Active INFORM. In Proceedings of the 14th International Symposium on Magnetic Bearings (ISMB14), Linz, Austria, 11–14 August 2014.
13. Schöb, R.; Redemann, C.; Gempp, T. Radial Active Magnetic Bearing for Operation with a 3-Phase Power Converter. In Proceedings of the ISMST4, Gifu City, Japan, 30 October–1 November 1997; pp. 111–124.
14. Ahn, H.J.; Jeong, S.N. Driving an AMB system using a 2D space vector modulation of three-leg voltage source converters. *J. Mech. Sci. Technol.* **2011**, *25*, 239–246. [CrossRef]
15. Hofer, M.; Hutterer, M.; Schrödl, M. PCB Integrated Differential Current Slope Measurement for Position-Sensorless Controlled Radial Active Magnetic Bearings. In Proceedings of the 15th International Symposium on Magnetic Bearings (ISMB15), Kitakyushu, Japan, 3–6 August 2016.
16. Hofer, M.; Wimmer, D.; Schrödl, M. Analysis of a Current Biased Eight-Pole Radial Active Magnetic Bearing Regarding Self-Sensing. In Proceedings of the 16th International Symposium on Magnetic Bearings (ISMB16), Beijing, China, 13–17 August 2018.
17. Herzog, R.; Vullioud, S.; Amstad, R.; Galdo, G.; Muellhaupt, P.; Longchamp, R. Self-Sensing of Non-Laminated Axial Magnetic Bearings: Modelling and Validation. *J. Syst. Des. Dyn.* **2009**, *3*, 443–452. [CrossRef]
18. Hofer, M.; Hutterer, M.; Schrödl, M. Application of Soft Magnetic Composites (SMCs) in Position-Sensorless Controlled Radial Active Magnetic Bearings. In Proceedings of the 15th International Symposium on Magnetic Bearings (ISMB15), Kitakyushu, Japan, 3–6 August 2016.

19. Gua, Y.; Zhu, J.G. Application of Soft Magnetic Composite Materials in Electrical Machines. *Aust. J. Electr. Electron. Eng.* **2006**, *3*, 37–46. [CrossRef]
20. Van der Broeck, H.W.; Skudelny, H.-C.; Stanke, G.V. Analysis and Realization of a Pulsewidth Modulator Based on Voltage Space Vectors. *IEEE Trans. Ind. Appl.* **1988**, *24*, 142–150. [CrossRef]
21. Chattopadhyay, S.; Mitra, M.; Sengupta, S. *Electric Power Quality*; Springer: Berlin, Germany, 2011; pp. 89–96, ISBN 978-94-007-0634-7.
22. Quang, N.P.; Dittrich, J.A. *Vector Control of Three-Phase AC Machines: System Development in Practice*; Springer: Berlin, Germany, 2015; pp. 17–59, ISBN 978-3-662-46915-6.
23. Hutterer, M.; Schrödl, M. Control of Active Magnetic Bearings in Turbomolecular Pumps for Rotors with Low Resonance Frequencies of the Blade Wheel. *Lubricants* **2017**, *5*, 26. [CrossRef]
24. Nerg, J.; Pöllänen, R.; Pyrhönen, J. Modelling the Force versus Current Characteristics, Linearized Parameters and Dynamic Inductance of Radial Active Magnetic Bearings Using Different Numerical Calculation Methods. *WSEAS Trans. Circuits Syst.* **2005**, *4*, 551–559.
25. Maslen, E.; Montie, D.; Iwasaki, T. Robustness Of Self Sensing Magnetic Bearings Using Amplifier Switching Ripple. In Proceedings of the 9th International Symposium on Magnetic Bearings (ISMB9), Lexington, KY, USA, 3–6 August 2004.
26. Hutterer, M.; Hofer, M.; Schrödl, M. New Results on the Robustness of Selfsensing Magnetic Bearings. In Proceedings of the 15th International Symposium on Magnetic Bearings (ISMB15), Kitakyushu, Japan, 3–6 August 2016.

 © 2019 by the authors. Licensee MDPI, Basel, Switzerland. This article is an open access article distributed under the terms and conditions of the Creative Commons Attribution (CC BY) license (http://creativecommons.org/licenses/by/4.0/).

Article

Thermal Behavior of a Magnetically Levitated Spindle for Fatigue Testing of Fiber Reinforced Plastic

Daniel Franz *, Maximilian Schneider, Michael Richter and Stephan Rinderknecht

Institute for Mechatronic Systems in Mechanical Engineering, Technische Universität Darmstadt, 64287 Darmstadt, Germany; schneider@ims.tu-darmstadt.de (M.S.); richter@ims.tu-darmstadt.de (M.R.); rinderknecht@ims.tu-darmstadt.de (S.R.)

* Correspondence: franz@ims.tu-darmstadt.de

† This paper is an extended version of our paper published in: Franz, D.; Schneider, M.; Richter, M.; Rinderknecht, S. Magnetically levitated spindle for long term testing of fiber reinforced plastic. In Proceedings of the 16th International Symposium on Magnetic Bearings (ISMB16), Beijing, China, 13–17 August 2018.

Received: 3 April 2019; Accepted: 30 April 2019; Published: 3 May 2019

Abstract: This article discusses the critical thermal behavior of a magnetically levitated spindle for fatigue testing of cylinders made of fiber reinforced plastic. These cylinders represent the outer-rotor of a kinetic energy storage. The system operates under vacuum conditions. Hence, even small power losses in the rotor can lead to a high rotor temperature. To find the most effective way to keep the rotor temperature under a critical limit in the existing system, first, transient electromagnetic finite element simulations are evaluated for the active magnetic bearings and the electric machine. Using these simulations, the power losses of the active components in the rotor can be derived. Second, a finite element simulation characterizes the thermal behavior of the rotor. Using the power losses calculated in the electromagnetic simulation, the thermal simulation provides the temperature of the rotor. These results are compared with measurements from an experimental spindle. One effective way to reduce rotational losses without major changes in the hardware is to reduce the bias current of the magnetic bearings. Since this also changes the characteristics of the magnetic bearings, the dynamic behavior of the rotor is also considered.

Keywords: active magnetic bearings; kinetic energy storage; fiber reinforced plastic; fatigue testing; thermal behavior

---

1. Introduction

Flywheels store energy as kinetic energy of the rotor and can provide a cost efficient solution for short-term energy storage and load smoothening services in electricity grids (e.g., [1,2]). Their advantages lie in the high possible number of cycles and the low initial cost per unit of power. The main drawbacks are their high stand-by-losses and the relatively low energy density compared to other storage technologies. In order to utilize the advantages, the energy density of one system should be increased while decreasing the stand-by losses at the same time.

1.1. Outer-Rotor Fywheel Design

One possible flywheel design is an outer-rotor setup (e.g., [3–5]). It promises high energy densities due to the large radii of the rotor and high rotational speeds. A realized full-scale system is described in [6]. A model of the system is shown in Figure 1a. The rotor is a magnetically levitated hollow cylinder. The outside of the rotor is made out of fiber reinforced plastic (FRP) with a circumferential fiber orientation, using carbon fibers. The inside of the rotor consists of different rotor parts of the active and passive components. Active magnetic bearings (AMBs) are used for radial levitation and a passive magnetic bearing is used for axial levitation. A permanent magnet synchronous machine (PMSM) accelerates and decelerates the rotor. All rotating components of the magnetic bearings and the PMSM are integrated on the inner circumference of the FRP rotor. Under rotation, these components press against the FRP, resulting in radial compressive stress transversal to the fiber orientation, which is superimposed by circumferential stress in the fiber direction (see Figure 1b). The energy density of the system increases with the radii of the rotor and its rotational speed, both factors lead to an increased stress in the FRP [7]. Consequently, high stress in the material is necessary to reach high energy densities. Charging and discharging the flywheel leads to cyclically varying mechanical stresses. To investigate the cyclic stability and lifetime of the FRP rotor, cyclic material tests, such as cyclic transverse tensile tests (derived from [8]), cyclic transverse compressive tests (derived from [9]), and cyclic four-point bending tests are performed on material samples. These samples are thin walled and relatively easy to fabricate with a high quality, whereas the rotor of the flywheel is much thicker and harder to produce in an industrial fiber winding process. Furthermore, the thermal expansion of carbon fiber and the epoxy plastic matrix used differ, leading to inner stress when a thick walled FRP structure cools down from the curing temperature. Hence, there is a much higher probability of imperfections and defects in the thick walled FRP structure of an outer-rotor flywheel than in the thin material samples. Therefore, the applicability of test results obtained with thin material samples to the rotor of the flywheel has to be investigated by tests with thick walled rotors. A specialized test rig was designed and set up to perform cyclic tests on thick-walled FRP specimens, which represent the outer-rotor of kinetic energy storage. These specimens are smaller, cheaper and have a lower energy content, reducing the danger during destructive testing.

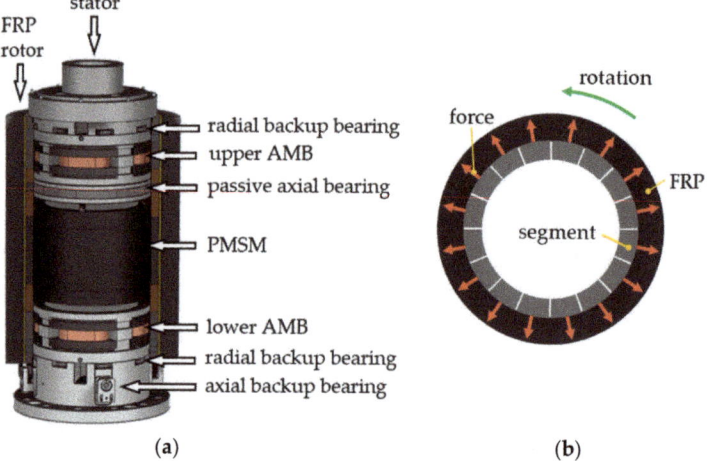

Figure 1. (a) Model the outer-rotor Flywheel described in [6] with a halved rotor; (b) Top view on the rotor showing the segments pressing against the fiber reinforced plastic (FRP) under rotation.

1.2. Testing Procedure

The ratio between the circumferential and the radial stress in a test specimen should be as close as possible to the one present in a full-scale flywheel. In the system described in [6], the radial transverse compressive stress in the FRP is 60 MPa and the circumferential longitudinal stress is 389 MPa at its maximum speed of 15,000 rpm. In the design of the flywheel, the maximum stress was limited to half of the expected strength of the FRP, to account for the high data uncertainty. To test the limits of the FRP, the stress in the specimen is doubled compared to the flywheel. To create this state of stress in the test specimen, a circumferentially segmented steel ring is placed inside the FRP cylinder of the specimen, and rotated at 30,000 rpm. This results in a transverse compressive stress of 100 MPa and a longitudinal tension of 775 MPa in the FRP. With an outer diameter of 190 mm, the surface speed of the specimen reaches approximately 298 m/s. The fatigue test will be conducted by alternating between a maximum speed of 30,000 rpm and a minimum speed of 15,000 rpm. At minimum speed the transversal compressive stress is 26 MPa and the longitudinal tension is 200 MPa. Because of the well-known high tensile strength of carbon fiber, the chance of fiber fracture is low at this state of stress. The probability of matrix fracture is expected to be much higher [10]. It is planned to perform up to 200,000 cycles per specimen. With a cycle time of 30 s this will take about 70 days. Additionally, overload tests are planned, where the specimen is accelerated to 40,000 rpm, resulting in a surface speed of nearly 398 m/s, a circumferential longitudinal tension of 1,400 MPa and a radial compressive stress of 180 MPa. The latter should lead to failure of the matrix. In the longitudinal direction, the FRP can withstand static tensions over 2,000 MPa. To reduce air drag and the consequential heating of the specimen, all tests are performed under vacuum conditions, with a pressure below 0.05 Pa.

1.3. Test Rig Description

Figure 2a shows a section view of the designed test rig, which was introduced in [11]. A hub and a shaft coupling connected the specimen to the driving spindle. This configuration, rather than an outer-rotor setup, was chosen to protect the active parts of the spindle in case of a failure of the specimen at high speed. Each of the eight segments of the steel ring inside the FRP weighed 1.027 kg, and the distance between their center of gravity (CoG) and the rotation axis is about 55 mm. Hence, at 40,000 rpm one segment contained around 30 kJ of kinetic energy. The surrounding containment was designed to absorb this energy in case of specimen failure. The design criteria for the containment were derived from [12]. The containment also serves as a vacuum chamber. A detailed view of the spindle is shown in Figure 2b. The motor in the middle is a water cooled PMSM with four poles, a maximum torque of 9.6 Nm, and a maximum power of 30 kW. In order to avoid excessive wear of the high-speed drive, the rotor is supported by AMBs. The radial AMBs were designed in a heteropolar configuration with laminated cores on rotor and stator, both made out of 0.2 mm thick sheets of non-oriented silicon steel (NO20). The radial position of the rotor is measured with four eddy current sensors at each bearing, two for each radial direction.

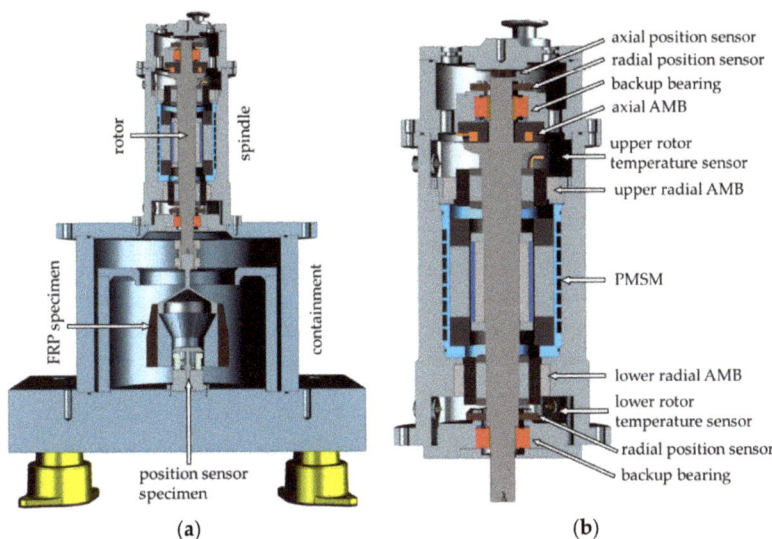

Figure 2. (a) Cross section of the test rig; (b) Cross section of the spindle.

To lift the combined mass of the spindle rotor and the specimen of roughly 20 kg, the axial AMB needs a large pole area, which leads to a big outer diameter of the thrust disk. Because of the high stress at high speed, a monolithic design was chosen (see [13]). The stator of the axial AMB was fabricated of the soft magnetic composite Siron S400b of the PMG Füssen GmbH [14]. An inductive sensor at the upper end of the rotor detects the axial position. Two-rowed hybrid spindle bearings with a static load rating of 20 kN are used as backup bearings. The rotor temperature is measured with two infrared sensors—one between the upper radial AMB and the thrust disk and one next to the lower radial position sensors, below the lower AMB. The moment of inertia of the specimen is about ten times higher than that of the rotor, and the surface speed of the specimen is more than two times higher. To protect the rotor and the spindle in case of a specimen failure, a predetermined breaking point was included in the hub (see Figure 3). This tapering also uncouples the dynamics of the rotor from the specimen to some extent.

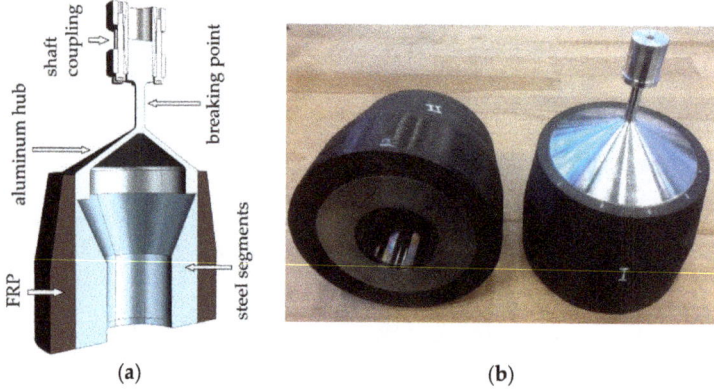

Figure 3. (a) Cross section of the specimen; (b) Assembled specimens.

One goal of the test rig operation is to minimize the total test duration for one specimen, maximizing the number of specimens that can be tested in a given amount of time. The test rig should thus be operated with as much power as possible. High power normally leads to high losses, as well as subsequent heating. Losses on the stator can be effectively cooled with water, but without a medium for convection or thermal conduction through physical contact, the rotor can only transfer heat to the stator via radiation. Hence, even small power losses in the rotor can lead to a high rotor temperature. The temperature of the rotor is crucial for the feasibility of the cyclic fatigue testing, since the temperature of the specimen affects the strength of the FRP and can change the test result. Furthermore, the magnets of the PMSM should not be operated over a certain temperature. Thus, the overall goal of this paper is to minimize the test duration under the boundary condition of a maximum rotor and specimen temperature. For this purpose, the losses in the rotor during operation and their influence on the rotor temperature have to be identified. Furthermore, feasible methods of minimizing the most crucial losses in an existing test rig are derived.

2. Method

With the goal of finding the shortest total test duration, this paper proposes a structured approach to evaluate the heating of a magnetically levitated rotor in a vacuum, and deduces measures to avoid critical temperatures. An overview on the approach is shown in Figure 4. First, losses in the rotor of both radial AMBs, the axial AMB, and the PMSM were calculated for different rotational speeds, using transient electromagnetic 2D finite element (FE) models implemented in ANSYS Maxwell 2015. The highest and lowest operating speeds of the spindle followed from the demanded stress in the FRP, as described in Section 1.2. For each component, a polynomial was fitted to the related losses, to obtain a mean loss value, which was used for the calculation of the steady state temperature. Next, air friction losses were calculated analytically, based on the kinetic theory of gases. Finally, to calculate the rotor temperature, a 3D-FE-model of a quarter of the test rig was set up and evaluated using ANSYS Workbench 19.1. Because of the water cooling, losses on the stator were not considered; instead the stator temperature in the simulation was predefined with measured data. A transient evaluation of the model was compared to measured temperatures on the test rig, to adjust thermal and loss calculation. To compute the rotor temperature after a long time of cycling, a steady state evaluation of the adjusted model was used. From this model, the most influential components on the rotor temperature can be derived. The focus was on components that allow for an improvement of the thermal behavior without major changes in the hardware. One option, which will be discussed as an example, is the reduction of the bias current in the radial AMBs [15]. Since this also changed the behavior of the rotor in the AMBs, a brief analysis of the rotor dynamics was performed using measurements and a mechanical 3D-FE-model of the rotor implemented in ANSYS Workbench 19.1. Another loss reduction can be achieved by reducing the control activity of the AMB [15]. Most of these strategies, like unbalance compensation (e.g., [16]) or a linear-quadratic-Gaussian-control (e.g., [17]), were not feasible at this point due to limitations of the AMB controller hardware. If no further loss reduction can be achieved in the components, the next step is to adjust the cycle time. Since rotational losses increase with speed and losses in the PMSM also increase with acceleration, the cycle time has a major influence on the mean losses and therefore on the rotor temperature. If the rotor temperature is below the critical values, the cycle time can be reduced. If the critical values are exceeded, the cycle time has to be increased. These options will be discussed further on.

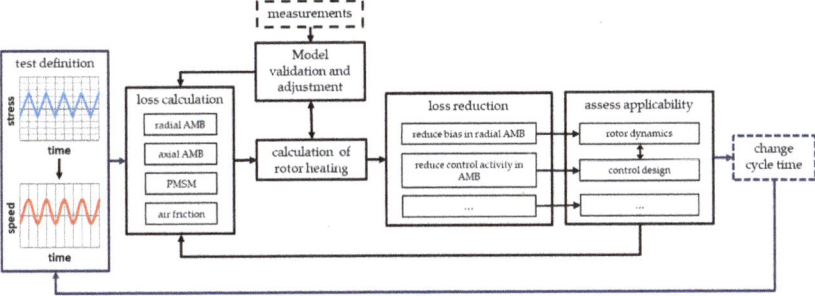

Figure 4. Applied approach to minimize the test duration by calculating and reducing rotor losses for the existing test rig.

3. Modeling of the Thermal Stability of the Rotor

The following sections discuss the loss calculation of the active magnetic bearings for the radial and axial direction and the electric machine, as well as losses caused by air friction. Subsequently, the thermal model focusing on the rotor temperature is described. For the model validation and potential adjustment, the calculation results are compared with measurements.

3.1. Loss Calculation of the Radial Active Magnetic Bearings

Both radial AMBs have eight poles, which are evenly distributed on the inner circumference of the stator and arranged in the sequence N–S–S–N–N–S–S–N. Flux leakage between two poles with equal flux direction—that is, N–N or S–S—is small, and thus two neighboring poles with differing flux direction, called a pole pair, function as one independent electromagnet. One pole pair of the upper radial AMB with an idealized average path of the magnetic flux φ is shown in Figure 5a. The magnetic flux of the AMB passes through the NO20-metal sheets that surrounded the solid rotor, which is shown in the bottom left corner. The nominal air gap between the rotor and the surrounding stator in the centered position is 0.4 mm. To linearize the actuator characteristic, a differential winding design is used, where a common bias coil and a control coil, with associated currents, are used for two counteracting pole pairs [18]. In the test rig, two coils are mounted on each pole of the radial AMBs, the inner for bias current I_B and the outer for the control current I_Y. Both coils have the same number of turns. All bias coils are connected in series for one AMB and the control coils for each direction in one AMB. The magnetic flux in a specific point in the rotor changes direction each time it passed a pole with a different flux direction. The flux also drops between two poles with equal flux direction. This remagnetization leads to hysteresis losses P_h and excess losses P_e, as well as eddy currents and subsequent losses P_c. To calculate the losses, a transient electromagnetic 2D-FE-model of the rotor was set up and evaluated. Figure 5b shows the simulated magnetic flux density B distribution in a quarter of the upper radial AMB at a rotational speed of 30,000 rpm. Only the bias current of $I_B = 5.67$ A magnetized the rotor; the control current was set to zero. The flux density lay at 0.7 T on average, but locally exceeded the saturation flux density of the used NO20 sheets of 1.1 T.

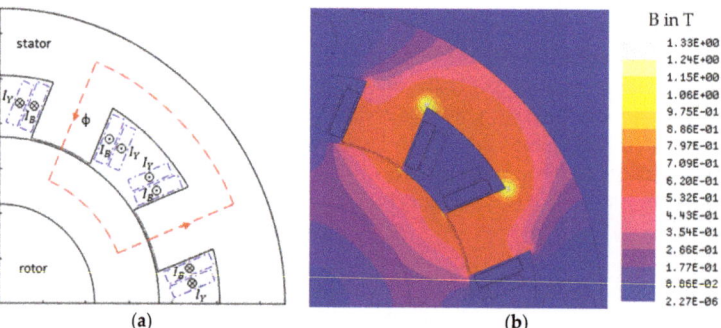

Figure 5. (a) Quarter of the cross section of the upper radial active magnetic bearings (AMB); (b) Magnetic flux density in the upper radial AMB with only the bias current as excitation. The rotor turns clockwise.

For the loss calculation, the Bertotti formula [19] in the form of Equation (1) was applied, in which the three loss mechanisms P_c, P_h, and P_e depend on the amplitude of the flux density B_m and the frequency f with which the magnetic flux changes:

$$P_B = P_c + P_h + P_e, \text{ with}$$
$$P_c = \int k_c (fB_m)^2 \, dV, \ P_h = \int k_h f B_m^2 \, dV, \text{ and } P_e = \int k_e (fB_m)^{1.5} \, dV. \tag{1}$$

The corresponding loss constants k_c, k_h, and k_e were derived by fitting data from the manufacturer of the electric steel obtained by standardized measurements [20] to Equation (1), which yields:

$$k_{c,AMB} = 0.23 \, \frac{\text{Ws}^2}{\text{T}^2 \text{m}^3} \text{ and } k_{h,AMB} = 193.6 \, \frac{\text{Ws}}{\text{T}^2 \text{m}^3}.$$

For the NO20 sheets used in the radial AMBs, $k_{e,AMB}$ could not be significantly determined and therefore was set to zero. The calculated total iron losses for the upper and lower radial AMB are shown in Table 1. Additionally, the switching of the amplifiers leads to further high frequency changes in the magnetic flux. The resulting rotor losses can be calculated with the same model and yielded 0.11 W for the upper AMB and 0.09 W for the lower AMB.

Table 1. Rotational losses in the radial AMBs.

	15,000 rpm	22,000 rpm	30,000 rpm
Upper AMB	7.6 W	14.3 W	24.1 W
Lower AMB	6.3 W	11.8 W	19.8 W

3.2. Loss Calculation the of Axial Active Magnetic Bearing

The magnetic flux in the axial AMB is mostly symmetrical to the rotational axis, so no significant remagnetization losses should occur due to rotation. Nevertheless, the switching of the amplifier leads to changes in the magnetic flux and associated rotor losses. Equation (1) is only applicable for thin sheets, but in the solid thrust disk of the axial AMB, without the usage of a laminated core or soft-magnetic composites, eddy current losses dominate. Therefore, the other loss mechanisms were neglected. To calculate eddy current losses a transient electromagnetic 2D-FE-model was analyzed. Figure 6a shows the magnetic flux density B in the axial AMB, with an air gap of 0.4 mm and a coil current of 3.15 A. The resulting force is 200 N, which is required to lift the rotor with the specimen, with a total weight of 20.3 kg. Changes in the magnetic flux in the rotor result in eddy currents and therefore in a non-zero current density J in the rotor. Eddy current losses $P_{c,J}$ were calculated as the integral of the square of J over the rotor volume V, divided by the conductivity of the material σ_{el} (see Equation (2)). The conductivity of X14CrMoS17 is $\sigma_{el,rot} = 1.43 \times 10^6 \frac{1}{\Omega m}$.

$$P_{c,J} = \frac{1}{\sigma_{el}} \int J^2 dV. \qquad (2)$$

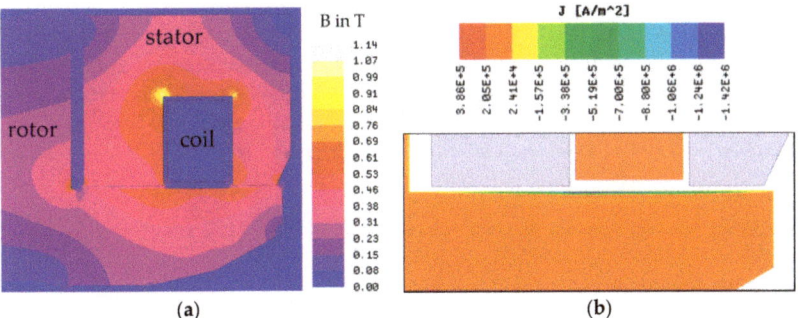

Figure 6. (a) Magnetic flux density in the axial active magnetic bearing (AMB); (b) Current density in the surface area of the thrust disk during switching of the axial AMB.

Since eddy currents oppose their provoking field, high frequency changes of the magnetic field only affect the outer layers of the conducting material. Hence, eddy currents are also limited to a thin region on the surface of the rotor (see Figure 6b). The losses of the axial AMB in the rotor due to the amplifier switching, derived by Equation (2), did not exceed 0.1 W.

3.3. Loss Calculation of the Permanent Magnet Synchronous Machine

The last active component that induces losses in the rotor is the PMSM. Here, the magnetic field rotates synchronously with the rotor. Remagnetization of the rotor occurs because of flux drops at slots in the stator and non-harmonic changes of the motor current due to switching of the inverter. Losses were calculated via a transient electromagnetic 2D-FE model. In the solid permanent magnets, losses were calculated using Equation (2), with an electric conductivity of the magnets of $\sigma_{el,mag} = 1.1 \times 10^6 \frac{1}{\Omega m}$. The rotor is laminated underneath the permanent magnets to further reduce losses. Remagnetization losses in these sheets were calculated according to Equation (1). The corresponding coefficients are:

$$k_{c,PMSM} = 0.12 \, \frac{Ws^2}{T^2 m^3}, \; k_{h,PMSM} = 166.7 \, \frac{Ws}{T^2 m^3}, \; \text{and} \; k_{e,PMSM} = 3.24 \, \frac{Ws^{1.5}}{T^{1.5} m^3}.$$

Results for different phase currents and rotational speeds are shown in Figure 7. The losses increased nearly linearly with the speed above 10,000 rpm. For root mean square values of the phase current (rms) below 10 A, losses were nearly independent of the phase current.

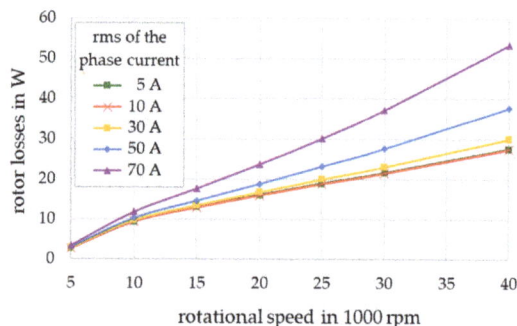

Figure 7. Rotor losses of the permanent magnet synchronous machine.

3.4. Air Friction Losses

For the calculation of air friction losses, it has to be evaluated if the gas in the system can be modeled with continuum dynamics. This was done by calculating the mean free path $\bar{l}$ of the gas. $\bar{l}$ is the avarage distance a molecule moves before it collides with another molecule of the gas. For a gas with pressure p, temperature T, and a mean atom diameter d_m, the mean free path $\bar{l}$ is calculated by Equation (3) (see [21]), where $k = 1.381 \times 10^{-23} \frac{J}{K}$ is the Boltzmann constant,

$$\bar{l} = \frac{kT}{\sqrt{2}\pi p d_m^2}. \tag{3}$$

The mean molecule diameter of air is $d_m = 3.559 \times 10^{-10}$ m [21]. During rotation, the maximum pressure in the test rig is $p = 0.05$ Pa and the minimum temperature $T = 290$ K, resulting in a minimum mean free path of $\bar{l} = 0.144$ m. Since $\bar{l}$ is much bigger than most of the air gaps in the test rig, an air molecule is more likely to collide with the walls of the test rig than with other air molecules. Hence, continuum dynamics are not applicable for this system. Instead, the kinetic theory of gases was utilized, where molecules are modeled as randomly moving elastic spheres. In [22], air friction losses P_a were derived by means of the momentum excange beween a spinning rotor and air molecules. For an axial disc with an outer radius r_o and an inner radius r_i this yields [22]:

$$P_{a,ax} = \pi p \sqrt{\frac{m_W \pi}{2kT}} \left(r_o^4 - r_i^4\right)\omega^2, \tag{4}$$

for a radial cylindrical surface with a radius r_r and a height h_r [22]:

$$P_{a,rad} = 4\pi p \sqrt{\frac{m_W \pi}{2kT}} h_r r_r^3 \omega^2, \tag{5}$$

and for a truncated cone with a height h_c, a lower radius r_l and an upper radius r_u:

$$P_{a,rad} = \pi p \sqrt{\frac{m_W \pi}{2kT}} \frac{\sqrt{(r_l - r_u)^2 + h_c^2}}{r_l - r_u} \left(r_l^4 - r_u^4\right)\omega^2, \tag{6}$$

where ω is the rotational speed of the rotor and m_W is the mean molecule mass of the gas. Hence, air friction losses increase linearly with p, quadratically with ω, and decrease with T. For dry air, the mean atom mass is approximately $m_W \approx 4.81 \times 10^{-26}$ kg [21]. The losses were calculated for each surface of the rotor and the specimen and were subsequently superimposed. For the worst case, with a maximum pressure of $p = 0.05$ Pa and a minimum temperature of $T = 290$ K, the sums of the calculated losses for the rotor and the specimen at different speeds are listed in Table 2.

Table 2. Air friction losses in the test rig with $p = 0.05$ Pa and $T = 290$ K.

	15,000 rpm	30,000 rpm
rotor	0.016 W	0.066 W
specimen	0.283 W	1.131 W

3.5. Simulation of Thermal Rotor Behavior

The thermal simulation was performed using a 3D-FE model of a quarter of the test rig, both with and without the specimen. The rotor transmits thermal energy to the stator via radiation, and vice versa. It was assumed that the emissivity and absorptivity of the rotor and stator surfaces do not depend on the wavelength or direction of the radiation. Under this assumption, the emissivity and absorptivity of the surface are equal [23]. For real materials, only a part of the radiation that hits a surface is absorbed while the rest is reflected, hence, the coefficient of emission is smaller than one. When considering the heat exchange between two surfaces through radiation, both surfaces have to be taken into account. Consequently, for the simulation of the rotor heating, the inner surface of the stator has to be included in the model. The stator and rotor were painted with lacquer on the surface of the AMBs and the PMSM. The coefficient of emission was assumed to be 0.9 for all painted parts and the FRP. For the unpainted, blank parts of the steel rotor, a coefficient of emission of 0.3 was assumed, and for the aluminum parts of the stator and the specimen a value of 0.05 was assumed [23]. The temporal stator temperature profile of the active components was roughly approximated with three measured temperatures, which were linearly interpolated. The temperature values can be found in Appendix A. This transient temperature change has only a small impact on the simulation results; using a constant mean temperature value for all stator components increases the calculated rotor temperature only about 1 °C. For the rest of the stator, a constant temperature of 25 °C was predefined. To obtain the temperature distribution in the rotor, heat conduction in the rotor was also taken into account in the model, with a conductivity of 25 W/Km. Calculating the steady state rotor temperature during the test cycles requires the mean loss value per cycle for each actuator. As a baseline, a cycling time of 30 s was defined—i.e., during one cycle, the rotor constantly accelerates for 15 s from 15,000 rpm to 30,000 rpm, and then constantly decelerates with the same slope back to 15,000 rpm. The rotational speed and the speed dependent losses of the active components are illustrated in Figure 8a. The mean values of all calculated losses are shown in Figure 8b.

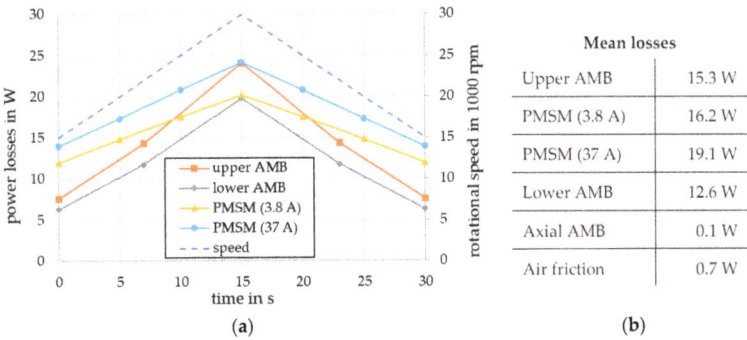

Figure 8. (a) Rotational speed and calculated rotational losses of the active components during one cycle; (b) Resulting mean losses of all calculated loss mechanisms in the rotor during cycling.

3.6. Temperature Measurements

To validate the described models, the rotor temperature was measured during cycling without a specimen. Two infrared sensors measured the rotor temperature: one between the upper radial AMB and the thrust disk and one next to the lower radial position sensors (see Figure 2b). The utilized sensors can measure the temperature of metallic surfaces above 50 °C, with an uncertainty of ±2 °C. First, the cooling of the levitated but not spinning rotor from a known temperature was analyzed. This excluded the rotational losses and the switching losses in the PMSM. Hence, only the small switching losses of the AMBs were present. The results of a transient simulation and the measurement are shown in Figure 9a. With a maximum deviation of 1.85 °C, the simulation results lie in the range of the measurement uncertainty. Consequently, the switching losses of the AMBs and the modeling of the heat transfer are sufficiently accurate. Next, the thermal behavior of the rotor during rotation was analyzed. Before the measurement, the system was turned off for 12 h, to ensure the rotor was cooled down to room temperature. During the measurement, the PMSM continuously accelerated and decelerated the rotor without a specimen from 15,000 rpm to 30,000 rpm. In the transient simulation, the mean losses from Figure 8b were used. The initial speed up to 15,000 rpm, which takes about 15 s, was neglected. Figure 9b compares the measured temperatures at both sensor positions with the simulation. While the rotor temperature at both positions was nearly the same in the simulation, it differed considerably in the measurement. The model underestimated the rotor heating at both measurement positions. At the lower position, the deviation was 5.2 °C, while it was 22.4 °C at the upper position. Due to the good agreement of the rotor cooling, the modeling error is expected to lie in the loss calculation rather than the thermal model.

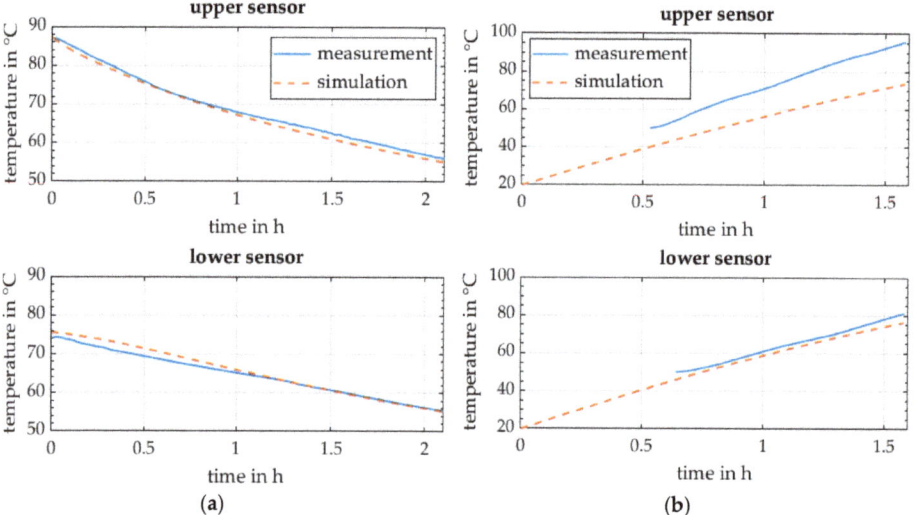

Figure 9. (a) Simulated and measured rotor cooling without the specimen; (b) Simulated and measured rotor temperatures without the specimen during cycling.

One reason why the losses could be underestimated can be found in the loss coefficients of the NO20 sheets in the AMBs and PMSM. The loss measurements done by the manufacturer that were used to identify the loss coefficients are normally performed on unmachined sheets after a precise annealing process. However, the material behavior can change during milling and heating. For example, a damaged sheet isolation can increase eddy current losses by up to 30% [24] and hysteresis losses can vary more than 50% depending on their heat treatment [25]. Because these changes can hardly be quantified, the calculated losses are adjusted to fit the temperature measurement. At the lower AMB, the calculated losses were increased by 10% to fit the temperature at the lower temperature sensor. Since the sheets at the upper and lower AMB were milled and heated similarly, the losses in the upper AMB increased by 10% as well. To further match the simulation with the measured temperature at the upper sensor, 8 W of additional losses in the axial AMB were required in the simulation. For the loss calculation, it was assumed that the axial AMB is perfectly symmetrical with respect to the rotational axis. In reality, small circumferential variations in the material of stator and rotor lead to flux changes during rotation and hence to higher rotor losses. Furthermore, control activities in the AMBs were neglected in the simulation, but during the measurement, considerable movement in the axial direction was observed. The adjusted losses are summarized in Table 3. A transient thermal simulation with the adjusted loss values showed good accordance with the measured data (see Figure 10a).

Table 3. Adjusted mean losses in the rotor during cycling.

Axial AMB	Upper AMB	Lower AMB
8.1 W	16.9 W	13.8 W

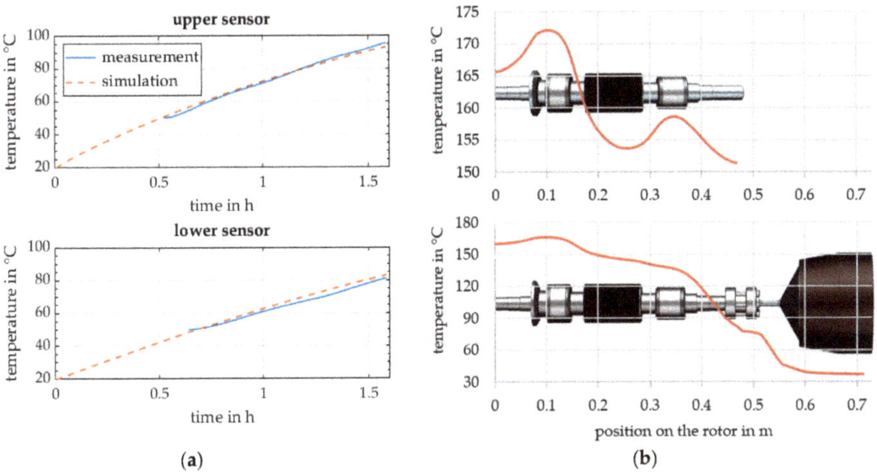

Figure 10. (a) Simulated and measured rotor temperatures with adjusted AMB losses without the specimen during cycling; (b) Simulated steady state temperature of the rotor with and without the specimen.

With the adjusted loss values, a steady state thermal simulation of the test rig with and without the specimen was performed. Without the specimen, the motor needs an rms phase current of 3.8 A to accelerate the rotor during the described cycle, but with the specimen an rms phase current of about 37 A is required. The losses of the PMSM were considered accordingly in the simulation. The other losses in the rotor were assumed to be equal for both cases. The simulated temperature distribution in the center of the rotor over the axial length is shown in Figure 10b. In the specimen, the temperature between the FRP and the steel segments is displayed. The maximum temperature without a specimen was 176 °C, located at the upper AMB. Even though higher rotor losses occur with the specimen, the rotor temperature was 10 °C lower than without it, as the large surface area of the FRP benefits heat transfer to the surrounding containment. The thin tapering at the hub led to a large temperature drop between rotor and specimen, preventing the temperature of the FRP from exceeding 40 °C. The operation temperature of the FRP in the flywheel must not exceed 80 °C. The magnets of the PMSM should not be operated over 120 °C. In order to have a safety margin, the magnet temperature should not exceed 110 °C. The simulation results, however, show a temperature of the magnets of over 140 °C. Thus, to avoid reducing the cycle time, rotor losses have to be reduced.

4. Loss Reduction

As seen in the previous section, the highest power losses in the rotor were due to rotational losses in the radial AMBs and PMSM, as well as switching losses in the PMSM, which all a have similar order of magnitude. Losses due to control activities in the axial AMB were slightly lower. Switching in the AMBs and air friction can be neglected. Hence, the main focus should be on the reduction of rotational losses in the radial AMBs and the PMSM as well as switching losses in the PMSM. In this paper only the radial AMBs will be further discussed.

4.1. Reducing Losses in the Radial Bearings

One effective way to reduce the rotational losses of an AMB is to reduce its bias current I_B. For example, reducing I_B from 5.67 A to 4 A reduces the rotational losses in the rotor by more than 50%, to 8.2 W in the upper AMB and 6.8 W in the lower AMB. A thermal simulation and measurement with $I_B = 4$ A was performed, where the rotor was again cycled according to Figure 8a. The results are shown in Figure 11. The transient simulation and the measurement of the test rig without a specimen again show sufficient correspondence (see Figure 11a). The steady state simulation predicted a maximum temperature of 129 °C at the axial AMB and 123 °C at the PMSM without a specimen, as well as 127 °C and 117 °C with a specimen (see Figure 11b). Changing I_B also changes the rotor dynamics in the AMBs. To evaluate the possibility of a bias reduction, the dynamics of the levitated rotor will be investigated in the next section.

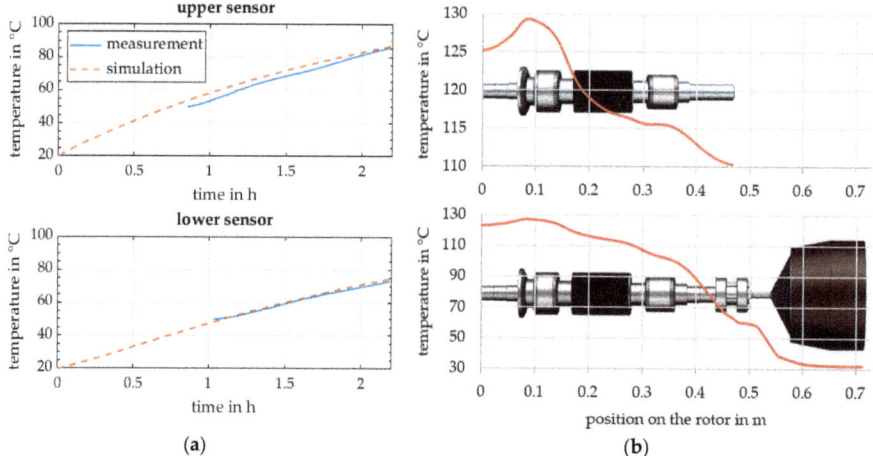

Figure 11. (a) Simulated and measured rotor temperatures with a reduced bias current of 4 A in both AMBs without the specimen during cycling; (b) Steady state rotor temperature with a reduced bias current of 4 A in both AMBs with and without the specimen.

4.2. Rotor Dynamics

A mechanical 3D-FE model of the rotor was created and evaluated. For a modal analysis of the model, the AMBs were modeled as springs with a constant stiffness of 2.5×10^5 N/m. The first bending mode of the rotor without a specimen lay at 1,030 Hz, thus 353 Hz above the highest rotational frequency. The calculated second bending mode was at 2,170 Hz. Measured Campbell diagrams of the rotor without the specimen are shown in Figure 12. They were derived from the signals of the radial position sensors of the AMBs during an acceleration of the rotor from 0 rpm to 40,000 rpm in 110 s. The signals were filtered with a 5 kHz low-pass filter and sampled with 10 kHz. To create the Campbell diagram, the measurement was divided into 4096 sample long windows, and for each window a Fourier transformation was performed. Figure 12a shows the Campbell diagram of the position data of the upper AMB with $I_B = 5.67$ A. The lines starting at the lower left corner are harmonics of the rotational speed, the thickest one being the first harmonic. Odd harmonics of the rotational speed are clearly visible, whereas even harmonics are barely detectable. The approximately horizontal lines are eigenfrequencies (EFs) of the system. At about 90 Hz lies the rigid body tilting mode of the rotor in the AMBs. The translational rigid body mode is not visible in the diagram. The first bending mode of the rotor can be seen at 1057 Hz, the second at 2063 Hz. The second splits into a forward and backward mode. While the calculated first bending mode only differed by 27 Hz, the second bending mode was about 100 Hz lower than the simulation predicted. Underneath each Campbell diagram, the mean deflection of the rotor for each window is shown. The deflections were less than 20% of the backup bearing clearance of 200 µm. Peaks can be seen when the rotational speed or its harmonics hit an EF. The first two peaks were due to the rigid body modes. The next peak at 25,000 rpm was caused by the fifth harmonic reaching the forward mode of the second bending EF, and the last peak at 39,000 rpm was caused by the third harmonic reaching the backward mode of the second bending EF. Smaller peaks can also be seen when higher harmonics reach the second bending EF, where every second odd harmonic excited its forward mode and every other odd harmonic excited the backward mode. A notch filter was placed at 1 kHz in the AMB control, so there was no additional excitation of this EF due to harmonics of the rotational speed. Figure 12b shows the same evaluations for $I_B = 4$ A. The harmonics were not influenced by I_B and the EFs only changed slightly. However, the amplitude of the deflection differed noticeable. The peak where the first harmonic crossed the second rigid body mode was about 30% higher with a reduced I_B, whereas the peaks at higher speeds were now below 20 µm. Because the normal speed range starts at 15,000 rpm, the rigid body mode has to be crossed only once per test, while the peak at 25,000 rpm is reached twice every cycle. Consequently, without a specimen, reducing I_B does not only reduce the rotational losses, but also reduces the control activity of the AMB during cycling.

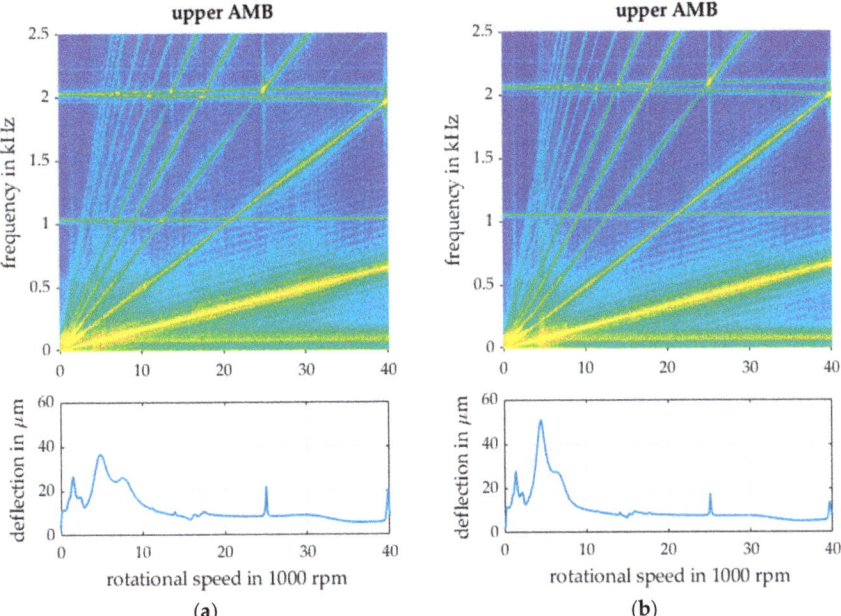

Figure 12. (a) Measured Campbell diagram and deflection of the rotor with a bias current of 5.67 A without specimen; (b) Measured Campbell diagram and deflection of the rotor with a reduced bias current of 4 A without specimen.

Measurements with a specimen were not performed yet, but Figure 13a shows the calculated EFs and Figure 13b the simulated Campbell diagram of the rotor with the specimen. The first elastic mode was at 9 Hz, where only the specimen tilts at the tapering. At speeds higher than 9 Hz, the CoG of the specimen did not move translationally. In the second, third, and fourth modes, the specimen was tilting around its CoG and in the fifth mode it does not move at all. The rotor showed elastic deformations beginning with the fourth mode. Due to the high inertia ratio of the specimen, the higher EFs showed a strong dependency on the rotational speed, which can be seen in the Campbell diagram in Figure 13b. The first three EFs were below 250 Hz, where the fatigue testing begins, and have to be passed while starting a test. The fourth EF increased in such a way that it is not reached by speed-synchronous excitations within the operation range. However, due to the close proximity of the fourth mode, the bandwidth of the AMBs has to include this frequency. The fifth EF had a minimum distance of 290 Hz to the rotational speed and therefore it will have to be filtered out from the position signal.

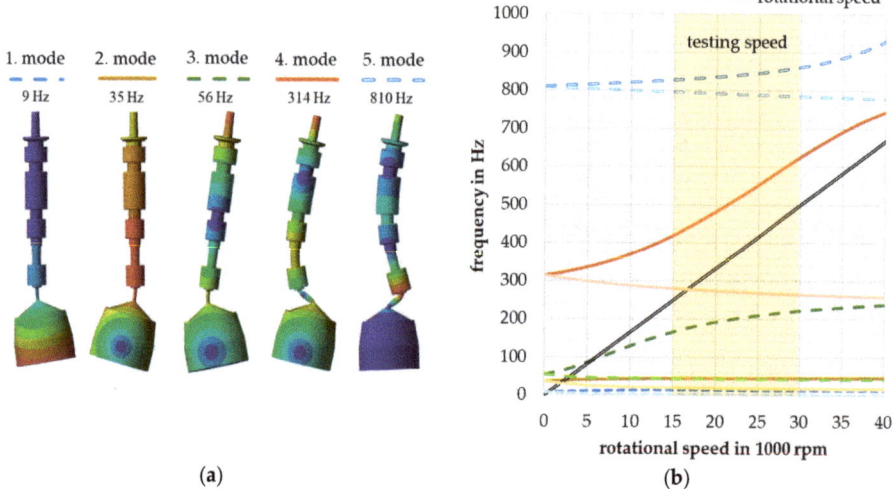

Figure 13. (a) Calculated bending eigenmodes below 1 kHz of the rotor with sample; (b) Calculated Campbell diagram of the rotor with the specimen. Each eigenmode shown in (a) splits in a forward and a backward mode.

For the differential winding design, which is used for the radial AMBs, the control current has to be smaller than I_B. To allow control currents bigger than I_B the AMBs would have to be rebuilt for a differential control design and a nonlinear control might be needed (see [15,26–28]). Both would require major changes in the hardware. Hence, with the existing hardware, reducing I_B also reduces the maximum force the actuators can generate. In the centered position, one AMB can generate a force of 278 N with $I_B = 5.67$ A, but only 144 N with $I_B = 4$ A. For the rotor without a specimen this is unproblematic, because there are nearly no external loads in the test rig and the rotor can be precisely balanced to minimize unbalance forces. However, the initial unbalance of an assembled specimen is big and balancing it precisely is difficult. The force generated by the AMB has to be big enough to counteract the unbalance force while passing the first three EFs. This has to be further investigated.

4.3. Increasing the Cycle Time

When no further loss reduction can be achieved, the cycle time has to be changed. In the upper example, even with a bias current of 4 A, the rotor temperature were above 110 °C but below 120 °C, as shown in Section 4.1. If further bias reduction or other loss reduction strategies are not applicable, a less favorable possibility to reduce the rotor temperature is to increase the cycle time. Here, two variations were considered, which are illustrated in Figure 14a. In the first variation, the acceleration of the rotor is reduced to lower load dependent losses in the PMSM, in the second variation, after a cycle, the rotational speed is held constant at 15,000 rpm for a while to allow the rotor to cool down. How long the cycle has to be increased depends on the associated loss reduction and the maximum allowable rotor temperature. For example, the cycle time could be increased from 30 s to 37 s to reduce the rms motor phase current from 37 A to 30 A, and thereby the rotor losses in the PMSM by 1.7 W to 17.4 W. Losses in the AMBs stay the same. Alternatively, after a 30 s cycle, the speed is held constant at 15,000 rpm for 7 s. Here, the mean losses can be calculated by averaging the mean losses for the first 30 s, as shown in Sections 3.5 and 3.6, and the losses at a constant speed. To keep the rotor without a specimen at a constant speed of 15,000 rpm, an rms phase current of 1.8 A is needed in the PMSM. Hence, it was assumed that an rms phase current of less than 10 A is needed to keep the rotor with a specimen at constant speed. Below 10 A, losses were nearly independent of the phase current (see Section 3.3). The resulting mean losses are summarized in Table 4. For the axial AMB constant, rotor losses of 8.1 W were assumed for both variations. The total cycle time in both variations was 37 s and the whole test would take the same amount of time—86 days to perform 200,000 cycles.

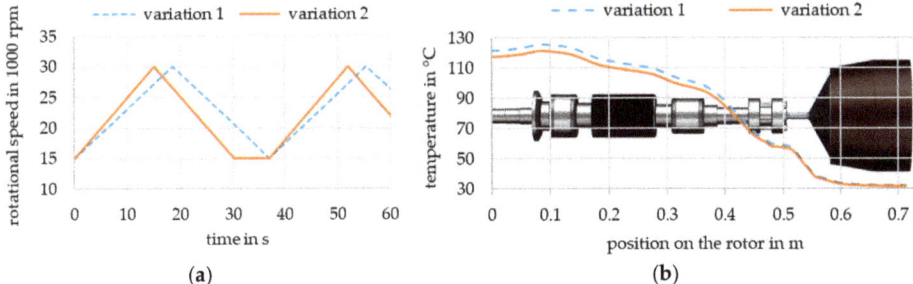

Figure 14. (a) Variations of increasing the cycle time to reduce the rotor temperature; (b) Steady state rotor temperature of both cycle variations with a reduced bias current of 4 A in both AMBs with a specimen.

Table 4. Mean rotor losses during cycling variation 1 and 2 with a reduced bias current of 4 A.

	Upper AMB	PMSM (30 A)	PMSM (37 A)	Lower AMB	Air Friction
Variation 1	8.2 W	17.4 W	-	6.8 W	0.70 W
Variation 2	7.5 W	-	17.7 W	6.2 W	0.59 W

Since losses in the PMSM showed only a small dependency on the load, the total mean losses in variation 2 were smaller than in variation 1. The calculated rotor temperature is shown in Figure 14b. For variation 1, the maximum calculated temperature in the permanent magnets of the PMSM was 115.6 °C, and for variation 2 it was 111.0 °C. Thus, with a slightly increased cooling time at 15.000 rpm, variation 2 should not reach the critical magnet temperature, whereas variation 1 has no big benefit for the rotor temperature. However, in the model the thermal conduction between rotor and magnets of the PMSM is assumed as ideal. In reality, small gaps could decrease the conductivity and therefore increase the dependency of the magnet temperature on the internal losses of the PMSM. This would make the loss reduction in the PMSM critical compared to the other components, giving variation 1 higher relevance.

5. Discussion

This paper proposed a structured approach to evaluate the heating of a magnetically levitated rotor in a vacuum, and deduced measures to avoid critical temperatures. First losses and the subsequent rotor heating were calculated. The rotational losses in the radial AMBs and the PMSM were identified as the main rotor loss mechanisms in the test rig. Losses in the axial AMB during rotation are smaller but also in the same order of magnitude. Switching losses in the AMBs and air friction losses only have a very small contribution to the total losses. One way to achieve loss reduction in the radial AMBs, without major changes in the spindle hardware, is to reduce the bias current. While doing so, the controllability of the rotor has to be ensured. The applicability of this approach was shown for the rotor without a specimen, but was not yet shown with a specimen. Stability problems could arise at low rotational speeds where many EFs can be excited. An adjustment of the bias with the rotational speed might be necessary. Further loss reduction can be accomplished by reducing the control activity. A centralized control of the radial AMBs should reduce the activities in the upper radial AMB. To realize a centralized control, however, the amplifier needs to be replaced. An overview of loss reduction strategies in AMBs can be found, for example, in [15]. Losses in the PMSM can be reduced by increasing the switching frequency of its inverter or by employing an output filter. Both options also require changes in the hardware. If no further loss reduction can be achieved, the cycle time has to be adjusted, so that the critical rotor temperature is exactly reached. In the present case, the cycle time had to be increased. With a cycle time of 37 s and a bias current of 4 A in both radial AMBs, a calculated steady state temperature of 111 °C can be reached, however, this has the drawback of an increased test duration. To account for modeling errors, the cycle time can further be adjusted during operation, depending on the rotor temperature, in a closed control loop. For the calculations, a constant thermal conductivity in the rotor was assumed. This assumption is especially critical between the magnets and the rotor. In reality, the thermal coupling will be poorer, resulting in a higher magnet temperature.

Author Contributions: Conceptualization, D.F., M.S. and M.R.; methodology, D.F., M.S.; validation, D.F.; formal analysis, D.F.; investigation, D.F.; resources, D.F.; data curation, D.F.; writing—original draft preparation, D.F.; writing—review and editing, D.F., M.S., M.R. and S.R.; visualization, D.F.; supervision, S.R.; project administration, D.F. and S.R.

Funding: This research was funded by the German Federal Ministry for Economic Affairs and Energy, grant number 03ET6064A.

Conflicts of Interest: The authors declare no conflict of interest. The funders had no role in the design of the study; in the collection, analyses, or interpretation of data; in the writing of the manuscript, or in the decision to publish the results.

Appendix A

The following Table A1 shows the measured stator temperature of the test rig during cool down. Table A2 shows the same temperatures during cycling with a bias current of 5.67 A and Table A3 shows them during cycling with a bias current of 4 A. These temperatures were used in their corresponding simulations.

Table A1. Measured stator temperature during cool down.

Time in s	axial AMB	Time in s	Upper AMB	Lower AMB	Time in s	Motor
0	25.0 °C	0	38.8 °C	40.9 °C	0	35.4 °C
4500	23.1 °C	2500	35.7 °C	37.3 °C	550	23.3 °C
7600	22.5 °C	7600	34.6 °C	35.6 °C	7600	22.3 °C

Table A2. Measured stator temperature during cycling with a bias current of 5.67 A.

Time in s	Axial AMB	Time in s	Upper AMB	Lower AMB	Time in s	Motor
0	18.6 °C	0	18.9 °C	18.9 °C	0	20.7 °C
1800	22.4 °C	1600	43.0 °C	45.3 °C	800	35.2 °C
5720	25.9 °C	5720	51.1 °C	53.3 °C	5720	35.5 °C

Table A3. Measured stator temperature during cycling with a bias current of 4 A.

Time in s	Axial AMB	Time in s	Upper AMB	Lower AMB	Time in s	Motor
0	18.5 °C	0	18.9 °C	19.0 °C	0	20.2 °C
2800	23.0 °C	1850	32.7 °C	34.9 °C	800	35.2 °C
8000	25.0 °C	8000	38.8 °C	40.9 °C	8000	35.4 °C

References

1. Sebastián, R.; Peña-Alzola, R. Flywheel energy storage systems: Review and simulation for an isolated wind power system. *Renew. Sustain. Energy Rev.* **2012**, *16*, 6803–6813. [CrossRef]
2. Amiryar, M.E.; Pullen, K.R. A Review of flywheel energy storage system technologies and their applications. *Appl. Sci.* **2017**, *7*, 286. [CrossRef]
3. Mason, P.; Howe, D.; Atallah, K. Soft Magnetic Composites in Active Magnetic Bearings. In Proceedings of the IEE Colloquium on New Magnetic Materials—Bonded Iron, Lamination Steels, Sintered Iron and Permanent Magnets, London, UK, 28 May 1998. [CrossRef]
4. Pichot, M.A.; Driga, M.D. Loss reduction strategies in design of magnetic bearing actuators for vehicle applications. *IEEE Trans. Magn.* **2005**, *41*, 492–496. [CrossRef]
5. Schaede, H.; Richter, M.; Quurck, L.; Rinderknecht, S. Losses in an Outer-Rotor-Type Kinetic Energy Storage System in Active Magnetic Bearings. In Proceedings of the 14th International Symposium on Magnetic Bearings, Linz, Austria, 11–14 August 2014.
6. Quurck, L.; Richter, M.; Schneider, M.; Franz, D.; Rinderknecht, S. Design and practical realization of an innovative flywheel concept for industrial applications. In Proceedings of the SIRM 2017–12 Internationale Tagung Schwingungen in Rotierenden Maschinen, Graz, Austria, 15–17 February 2017; pp. 151–160. [CrossRef]
7. Arnold, S.M.; Saleeb, A.F.; Al-Zoubi, N.R. Deformation and life analysis of composite flywheel disk systems. *Compos. Part B Eng.* **2002**, *33*, 433–459. [CrossRef]
8. ISO 527-5, Plastics—Determination of Tensile properties—Part 5: Test Conditions for Unidirectional Fibre-Reinforced Plastic Composites. Available online: http://www.iso.org/standard/52991.html (accessed on 2 May 2019).
9. ISO 14126, Fibre-Reinforced Plastic Composites—Determination of Compressive Properties in the in-Plane Direction. Available online: https://www.iso.org/standard/23638.html (accessed on 2 May 2019).
10. Knops, M. *Analysis of Failure in Fiber Polymer Laminates*; Springer: Berlin/Heidelberg, Germany, 2008; Volume 2. [CrossRef]
11. Franz, D.; Schneider, M.; Richter, M.; Rinderknecht, S. Magnetically levitated spindle for long term testing of fiber reinforced plastic. In Proceedings of the 16th International Symposium on Magnetic Bearings, Beijing, China, 13–17 August 2018.
12. Hagg, A.C.; Sankey, G.O. The containment of disk burst fragments by cylindrical shells. *ASME J. Eng. Power* **1974**, *96*, 114–123. [CrossRef]
13. Larsonneur, R. Design and Control of Active Magnetic Bearing Systems for High Speed Rotation. Doctoral dissertation, Swiss Federal Institute of Technology Zuerich, Zuerich, Switzerland, 1990. [CrossRef]
14. Schoppa, A.; Delarbre, P. Soft magnetic powder composites and potential applications in modern electric machines and devices. *IEEE Trans. Magn.* **2014**, *50*. [CrossRef]
15. Gaechter, S.; Kameno, H.J. Application of zero-bias current active magnetic bearings to flywheel energy storage systems. *Koyo Eng. J.* **2004**, *165*, 25–30.
16. Herzog, R.; Bühler, P.; Gähler, C.; Larsonneur, R. Unbalance compensation using generalized notch filters in the multivariable feedback of magnetic bearings. *IEEE Trans. Control Syst. Technol.* **1996**, *4*, 580–586. [CrossRef]
17. Barbaraci, G.; Pesch, A.H.; Sawicki, J.T. Experimental investigations of minimum power consumption optimal control for variable speed AMB rotor. In Proceedings of the ASME 2010 International Mechanical Engineering Congress & Exposition, Vancouver, BC, Canada, 12–18 November 2010. [CrossRef]
18. Schweitzer, G.; Maslen, E.H. *Magnetic Bearings*; Springer: Berlin/Heidelberg, Germany, 2009. [CrossRef]

19. Bertotti, G. General properties of power losses in soft ferromagnetic materials. *IEEE Trans. Magn.* **1988**, *24*, 621–630. [CrossRef]
20. IEC 60404-2, Magnetic Materials—Part 2: Methods of Measurement of the Magnetic Properties of Electrical Steel Strip and Sheet by Means of an Epstein Frame. Available online: https://webstore.iec.ch/publication/62746 (accessed on 2 May 2019).
21. Jousten, K. *Handbook of Vacuum Technology*, 2nd ed.; Wiley-VCH Verlag: Weinheim, Germany, 2012. [CrossRef]
22. Koch, R. Schwung-Energiespeicher-System mit Supraleitendem Magnetlager. Ph.D Thesis, University of Stuttgart, Stuttgart, Germany, 2002.
23. Baehr, D.B.; Stephan, K. *Heat and Mass Transfer*, 3rd ed.; Springer: Berlin/Heidelberg, Germany, 2011. [CrossRef]
24. Müller, G.; Vogt, K.; Ponick, B. *Berechnung elektrischer Maschinen*, 6th ed.; Wiley-VCH Verlag: Weinheim, Germany, 2009. [CrossRef]
25. Canders, W.R. Berechnung von Eisenverlusten—Physikalisch Basierter Ansatz Nach Bertottis Theorie. Annual report, Braunschweig, Germany. 2011. Available online: https://www.tu-braunschweig.de/Medien-DB/imab/09-Jahresberichte/2010-11/07_Canders_2010_11.pdf (accessed on 2 May 2019).
26. Lévine, J.; Lottin, J.; Ponsart, J.C. A nonlinear approach to the control of magnetic bearings. *IEEE Trans. Control Syst. Technol.* **1996**, *4*. [CrossRef]
27. Wilson, B.C.; Panagiotis, T.; Heck-Ferri, B. Experimental validation of control designs for low-loss active magnetic bearings. In Proceedings of the AIAA Guidance, Navigation and Control Conference 2005, San Francisco, CA, USA, 15–18 August 2005. [CrossRef]
28. Grochmal, T.R.; Lynch, A.F. Nonlinear control of an active magnetic bearing with bias currents: experimental study. In Proceedings of the 2006 American Control Conference, Minneapolis, MN, USA, 14–16 June 2006. [CrossRef]

© 2019 by the authors. Licensee MDPI, Basel, Switzerland. This article is an open access article distributed under the terms and conditions of the Creative Commons Attribution (CC BY) license (http://creativecommons.org/licenses/by/4.0/).

Article

Design and Analysis of a 1D Actively Stabilized System with Viscoelastic Damping Support [†]

Josef Passenbrunner [1],*, Gerald Jungmayr [1] and Wolfgang Amrhein [2]

[1] Linz Center of Mechatronics GmbH, Altenbergerstr. 69, 4040 Linz, Austria; gerald.jungmayr@lcm.at
[2] Department of Electrical Drives and Power Electronics, Johannes Kepler University, Altenbergerstr. 69, 4040 Linz, Austria; wolfgang.amrhein@jku.at
* Correspondence: josef.passenbrunner@lcm.at; Tel.: +43-732-2468-6087
[†] This paper is an extended version of our paper published in Passenbrunner, J.; Jungmayr, G.; Amrhein W. Design of a Passively Magnetically Stabilized System with Viscoelastic Damping Support and Flexible Elements. In Proceedings of the 16th International Symposium on Magnetic Bearings (ISMB16), Beijing, China, 13–17 August 2018.

Received: 5 March 2019; Accepted: 15 April 2019; Published: 17 April 2019

Abstract: Passively magnetically stabilized degrees of freedom yield the benefit of reduced complexity and therefore costs. However, the application of passive magnetic bearings (PMBs) also features some drawbacks. The poor damping capability leads to exaggerated deflection amplitudes when passing the resonance speeds of the applied system. This results in the necessity of external damping. Complying with the goal of costs and complexity, viscoelastic materials offer a suitable solution. However, these materials show high frequency and temperature dependent properties which induce the necessity of a proper model. Thus, the design of systems, as presented in this paper, requires accurate modeling of the dynamic behavior including the nonlinear characteristic of damping elements to predict the system displacements. In the investigated setup only two degrees of freedom remain to be controlled actively. These are the axial rotation and the axial position of the rotor which are controlled by the motor and an active magnetic axial bearing (AMB). This article focuses on the rotor dynamic modeling of a radial passively magnetically stabilized system especially considering the nonlinear behavior of viscoelastic damping elements. Finally, the results from the analytic model are verified by measurements on a manufactures test system.

Keywords: passive magnetic bearing; viscoelastic material; damping; modeling; rotor dynamics; active magnetic bearing; cost reduction

1. Introduction

Nowadays, the number of applications for rotating machines is huge. In general, therefore, the rotor is supported with traditional mechanical ball or slide bearings, which feature the drawback of applied lubricants, mechanical friction and resulting wear. Magnetic bearing technology is a contemporary research area that can be described as relatively young. The principle is based on the generation of suspension forces by magnetic fields which allows a contact-free operation of the motor. Magnetic bearings have their advantages especially in areas that require high-purity operation or very high speeds. These applications can be found in machining technology as milling spindles, in vacuum technology as turbomolecular pumps, in medical applications as blood pumps as well as in flywheel accumulators [1,2] and gas compressors. Additionally, the aspect of service life is superior, due to the fact that in magnetically suspended systems it is only determined by the electronic components.

In addition, as is usual in the field of magnetically stabilized system design, the costs play a very important role. The high complexity compared to standard ball bearings has thereby a major impact on the price. With reference to fully actively stabilized systems, which are still applicable in commercial areas, investigations regarding passively stabilized systems are rare [3,4]. The first works on passive permanent magnetic (PM) ring bearings can be found in 1976 [5]. In 1981, Yonnet [6] characterized the different possible bearing principles for the first time. Marinescu and Yonnet presented the basics for dimensioning of PM ring bearings in 1979 and 1980 [7,8]. A great advance in analytical computability was presented by Lang [9] for attractive ring bearings. An extension to repulsive ring bearings was provided by Jungmayr [10]. Today the filed of applications for passive PM ring bearings include spinning centrifuges, turbomolecular pumps, flywheels and fully magnetically stabilized fans.

The use of passive magnetic bearings (PMB) offers a very interesting approach to reduce the complexity of the magnetic bearing to a minimum and thereby makes the application attractive for cost-sensitive areas. Hence, in such systems the demand for power electronics and position sensors is minimized. The passive bearing arrangement by means of permanent magnets has to be emphasized, as it enables a nearly loss-free bearing arrangement with very little effort. Furthermore it features the advantage of easy determination of stiffness and static load carrying capacity. The determination of the stiffness of permanent magnet bearings can be performed analytically in arrangements without ferromagnetic material. For applications with ferromagnetic material finite element calculations are necessary. However, a major disadvantage of permanent magnetic bearings is the extraordinary low damping of the stabilized degrees of freedom, which lead to exaggerating deflections when passing rigid body resonances during a run-up process. So additional damping has to be induced into the system [11–14]. Especially systems with large unbalances need a very precise consideration of the dynamic behavior.

A possibility to provide damping is offered by viscoelastic materials which meet the targets of low costs and simplicity. Viscoelastic materials are currently being used in various applications for damping and suppression of vibrations. The selection of the damping material itself is often empirical. One reason for this lies in the complex dynamic behavior of this type of materials. In most cases the stiffness and damping of elastomers feature a high dependency on the excitation frequency and temperature. Using viscoelastic materials it is no longer possible to actively influence the system dynamics subsequently. Thus, a precise consideration of the material characteristics is necessary. This implies an exact knowledge of the overall system, especially the rotordynamic behavior, to ensure proper operation. Occurring deflections, due to the rotor dynamics, have to meet certain restrictions, to avoid contact between the rotor and the stator. This requires the derivation of the systems equations of motion and a proper model of the frequency-dependent damping elements behavior.

The concept examined in this paper is based on an industrial application, which is located in the area of mass production. With a large number of manufactured systems the price plays a decisive role. Thus, only system concepts come into consideration, which can drastically reduce costs. The avoidance of actively stabilized degrees of freedom and the resulting hardware savings (electronics, sensor technology, bearing coils) are crucial. Concepts with passively magnetically stabilized degrees of freedom are favorable due to their simplicity and low costs. However, a solution can only be achieved if the above-mentioned problems of damping can be solved in a cost-effective way. Although the modelling of viscoelastic damping elements is very complex, it offers an approaches to set up a proper system description.

The aim of this work is the design and optimization of a low-cost magnetic bearing drive system taking into account the special boundary conditions of the application. The high reliability of the concept is to be demonstrated, helping to implement contactless bearings in the industrial field more often.

In Section 2 the system setup and components are described. Section 3 determines the viscoelastic material model followed by Section 4, where the derivation of the equations of motion is presented. With the derived model an optimization is performed in Section 5. Finally, the optimized system is verified by measurements in Section 6.

2. System Setup

Based on the application considered, the rotor features a vertical alignment. The speed range is 10.000–30.000 rpm, which, however, is constantly interrupted. Hence, a frequent start and stop of the system is demanded. A very critical point are unbalances, which will lead to high deflections in the passive bearings. These unbalances occur mainly due to constantly changing additional process masses, which are attached to the rotor during operation. Thereby the unbalance maximum is defined by a value of 20 gmm and can be applied to the rotor at any position. The drive system itself can be balanced in order to reduce the unbalance to a minimum. However, the unbalance of the process mass must be taken as the occurring unbalance. Especially in connection with a PMB concepts, this represents an immense challenge and requires an exact consideration of the damping elements.

Geometric restrictions exist on the outer diameter of the rotor which is given by a maximum value of 42 mm. This requires a very compact design of the system. The limited space conditions lead to a stacked vertical construction. Figure 1 shows the setup of the considered magnetically levitated system.

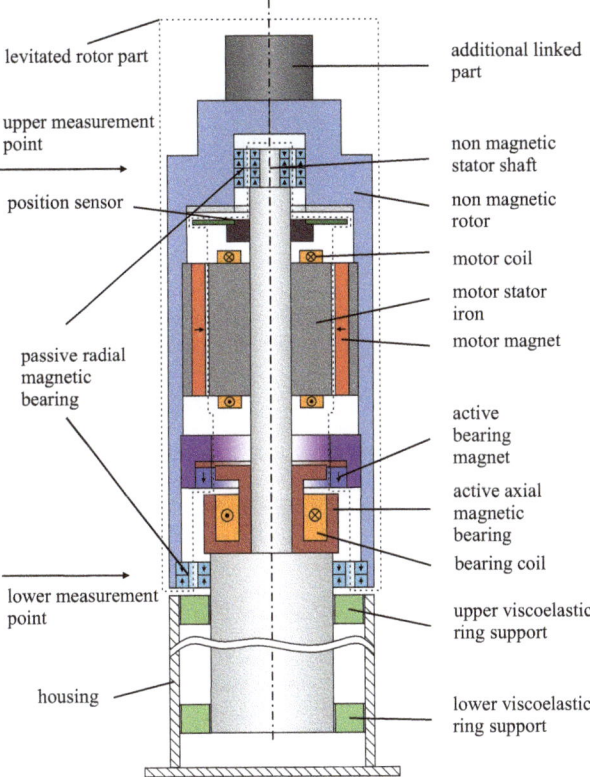

Figure 1. Setup of the investigated system.

In the shown configuration, the radial deflections and the tilting are stabilized by two radial permanent magnet bearings. These ringbearings are of repulsive type and use axial magnetized magnets, which are easy to produce and lower the costs for the bearings. Furthermore, this configuration offers the opportunity of magnet stacking [15] to achieve higher bearing stiffness with the same cross section and thereby simplifies the rotordynamic design procedure.

Without ferromagnetic material near the PM the stabilizing radial stiffness of the bearings can be analytically calculated [9]. This drastically simplifies the bearing design for the prototype, because no finite element simulations are necessary. Due to Earnshaw's theorem [16], which is adapted for passive ring bearings in [17], the destabilizing axial stiffness s_z of the bearings is given by

$$s_z = -2 \cdot s_r, \qquad (1)$$

where s_r describes the stabilizing radial stiffness.

Hence, at least one direction needs to be stabilized actively. In between the active axial bearing, the motor and the position sensor [18,19] are placed. As damping elements viscoelastic ring elements are used. These elements are located between the stator and the system housing. The viscoelastic ring elements are placed and glued in two aluminum rings to allow easy mounting. With this concept and the right dimensioning, it is possible to achieve improved dynamic system properties [20]. The rotor is designed as an exterior rotor. In principle, interior rotor concepts are also possible, but it should be avoided that the critical bending Eigenfrequencies of the rotor occur within the speed range. Moreover, the supporting rod of the motor, the active bearing and the sensor system become thin. Hence, the flexible behavior might as well affects the dynamics of the system. However, the considered motor is constructed with a slotted stator and ferrite magnets for cost reduction. As AMB a reluctance force bearing is used, whose force density is higher compared to a Lorenz force bearing.

3. Model of the Viscoelastic Behavior

Viscoelastic materials feature a frequency dependent behavior of the stiffness and damping values. To describe the characteristics of such materials often a generalized Maxwell model is used [21–23]. As shown in Figure 2, such a model consists of a single spring with the equilibrium modulus E_0 and several Maxwell units in parallel. This spring E_0 describes the material response after infinite time. Each Maxwell consists of a single spring and a single damper in series, reproducing the frequency dependency by adding different time constants τ_n. So a quasi non-linear behavior can be represented by superposing linear elements. For harmonic excitations in the frequency domain a representation of the generalized Maxwell model with a complex modulus

$$E(\omega) = E'(\omega) + jE''(\omega) \qquad (2)$$

is useful. Thereby, E' stands for the storage module and E'' for the loss module. The quotient gives the loss factor

$$\eta = tan\delta = \frac{E''}{E'}. \qquad (3)$$

Converted to the Prony parameters [24] of the generalized Maxwell model the components of the complex modulus result in

$$E'(\omega) = E_0 + \sum_{n=1}^{N} E_n \frac{\omega^2 \tau_n^2}{1 + \omega^2 \tau_n^2} \qquad (4)$$

and

$$E''(\omega) = \sum_{n=1}^{N} E_n \frac{\omega \tau_n}{1 + \omega^2 \tau_n^2} \qquad (5)$$

with the time constants $\tau_n = \frac{d_n}{E_n}$.

In addition, the Maxwell model uses N inner states y_n. A main advantage of this model is its easy integration as it can be described by a system of linear differential equations.

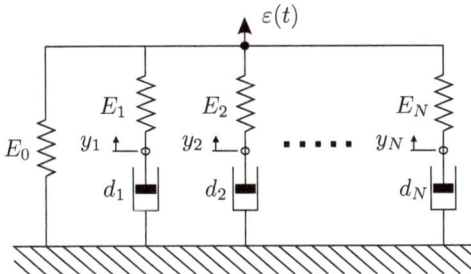

Figure 2. MAXWELL-Modell of the viscoelastic materials dynamic behavior representing one viscoelastic ring support.

Determination of Material Parameters

To describe the thermo-viscoelastic behavior usually master curves are used [24]. Thereby, the theory of temperature-time-correspondence is applied to combine measurements at different temperatures and draw conclusions for other frequencies. The basic approach is depicted in Figure 3.

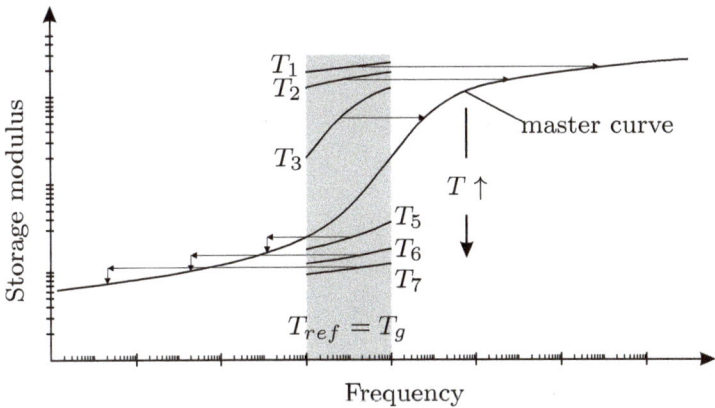

Figure 3. Application of the temperature-time-correspondence.

In theory [25] viscoelastic material shows the same storage modulus for a certain temperature T_1 and frequency f_1 as for a different temperature T_2 and a referring scaled frequency f_2 according

$$E(f_1, T_1) = E(f_1 \cdot a_{T_{2-1}}, T_2) = E(f_2, T_2) \quad (6)$$

with the shift factor $a_{T_{2-1}}$. With lower temperatures the storage modulus is increasing and falling with increasing temperatures. As only a limited frequency range can be measured, one measurement is not sufficient to represent the required frequency range. Hence, it is necessary to combine the measurements. The shift factors a_T thereby perform a vertical shift of the measured data, so that a smooth overall mastercurve is resulting. If unfilled elastomers are used, the kinetic-theory-factor [25] has to be applied, which is obtained from the theory of entropy elasticity. The storage modulus is thereby horizontally shifted to the reference temperature T_r using

$$E(T_r) = \begin{cases} E(T_m) * \underbrace{\frac{T_r}{T_m}}_{k_T} * \underbrace{\frac{\rho_r}{\rho}}_{\approx 1} & T_m \geq T_g \\ E(T_m) & T_m < T_g \end{cases} \quad (7)$$

with T_m as measured temperature and T_g as the glass transition temperature. The factor ρ_r/ρ normalizes the specific volume at a temperature T_m to the reference temperature T_r. Therefore, the horizontal shift is only applied to temperatures lower than the glass transition temperature, which lies for technical elastomers normally far below zero degrees Celsius. The loss factor values are also shifted in that process as it is the quotient of E' and E''.

Reliable master curve data is hardly provided by most manufacturers and even if available it is essential to know the exact measurement conditions. For example a measurement under precompression shows a very different behavior compared to the same measurement without precompression. Hence, it was decided to measure the data on our own. Even though the measurement principle looks simple, it is a very tricky task. In Figure 4 the employed measuring method is shown. It is based on the so called Dynamic-Mechanic-Temperature-Analysis (DMTA).

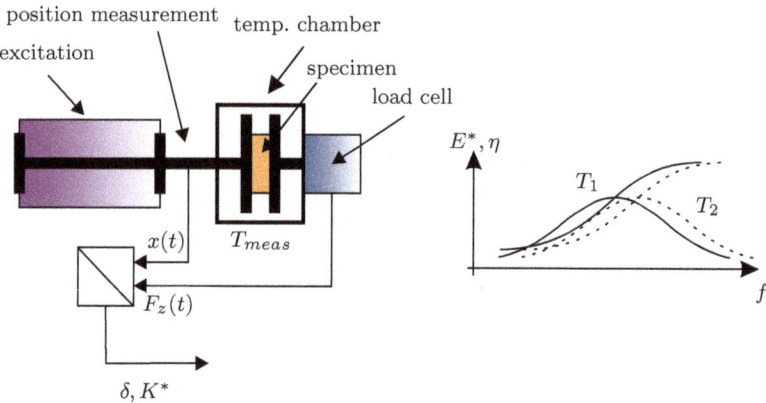

Figure 4. Scheme of the employed measuring method for the viscoelastic materials.

A specimen is excited at different temperatures by harmonic deformation $x(t)$ at various frequencies. In the process the reaction force $F(t)$ is measured by a load cell. Thereby, the ratio between force and displacement reflects the stiffness of the material. Due to dissipation effects a phase shift δ occurs between the force and the displacement, which represents the damping capability of the material. The measurement leads to the so-called isotherms. Thereafter, the measured stiffness values [26] have to be converted to the material significant storage modulus E' and loss modulus E'' using the geometric parameters of the specimen. For a circular specimen with the height h_s and diameter d_s the relation between the measured axial stiffness k_{ax} and the material modulus E is given by

$$k_{ax} = E \cdot \frac{A}{h_s} \cdot (1 + 2S^2) \quad S = \frac{d_s}{4h_s} \quad A = \frac{d_s^2 * \pi}{4}. \tag{8}$$

In Figure 5a,b the measured isotherms of a selected butyl rubber with a hardness of Shore A40 are shown. It can be seen that the storage modulus is increasing with lower temperature. However, the loss factor shows, at the beginning, an increase with lower temperature till it reaches a maximum value. Afterwards, the loss factor falls again.

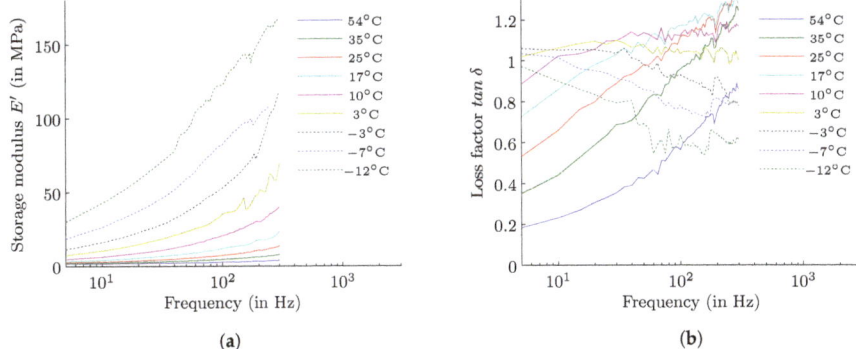

Figure 5. Storage modulus (**a**) and loss factor (**b**) of the measured isotherms.

For this material also measurement data from the manufacturer is available. It was observed that the measured data shows a 50% higher stiffness and a 30% lower loss factor compared to the given data provided by the manufacturer. That proves the difficulty to get proper material specifications. The isotherms are afterwards shifted using Equation (6) processing the vertical shift and (7) for the horizontal shift of the data below the glass transition temperature to obtain the master curve considering a smooth gradient of the stiffness and loss factor. However, there are other possibilities to shift the data. Basically the procedure to find the optimal shift parameters and consequently the identification of the Prony-parameters requires the solution of a nonlinear minimization problem. In this work a genetic algorithm is used varying the shift parameters and determineing the Prony-parameters by minimizing the mean square deviation. The results are plotted in Figure 6. The fitted model shows a very good compliance with the measured data.

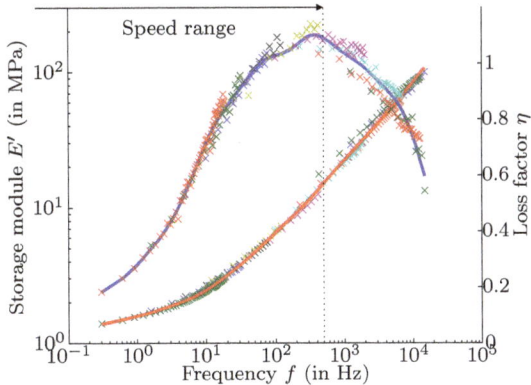

Figure 6. Fitted curves of the storage modulus (red) and the loss factor (blue) in comparison to the measured data (×) at a reference temperature of 25 °C, whereby the different colors mark the various measurement temperatures.

4. Rotordynamic Model

In the considered system, high rotor unbalances are induced by variable process masses acting on the rotor. Obviously, the relative deflection between stator and rotor in the PMB planes are important. To determine the movements of the system bodies the equations of motion have to be derived. Due to the system setup, the AMB, the motor and the upper PMB, are located on a thin shaft. Therefore, also the bending of this part have to be considered. The equations of motion are conducted by using the projection equation [27], which projects the (generalized) forces into the unconstrained space, where the motion takes place. These forces are given by

$$\sum_{i=1}^{N} \left[\left(\frac{\partial {}_R v_{si}}{\partial \dot{q}} \right)^T \left(\frac{\partial {}_R \omega_{si}}{\partial \dot{q}} \right)^T \right] \begin{bmatrix} ({}_R \dot{p} + {}_R \tilde{\omega}_{IR} \, {}_R p - {}_R f^s)_i \\ ({}_R \dot{L} + {}_R \tilde{\omega}_{IR} \, {}_R L - {}_R M^s)_i \end{bmatrix} + \left(\frac{\partial V}{\partial q} \right)^T + \left(\frac{\partial R}{\partial \dot{q}} \right)^T = 0, \quad (9)$$

with ${}_R v_{si}$ and ${}_R \omega_{si}$ describing the bodies velocities and angular velocities, while ${}_R p$ and ${}_R L$ represent the impulse and angular momentum in the reference system R. Thereby,

$$\left(\frac{\partial {}_R v_{si}}{\partial \dot{q}} \right)^T \quad \text{and} \quad \left(\frac{\partial {}_R \omega_{si}}{\partial \dot{q}} \right)^T \quad (10)$$

represent the Jacobian matrices and the therms

$$\left(\frac{\partial V}{\partial q} \right)^T \quad \text{and} \quad \left(\frac{\partial R}{\partial \dot{q}} \right)^T \quad (11)$$

are used to consider potential forces (e.g., springs, dampers, elastic potential and gravitation). ${}_R f^s$ and ${}_R M^s$ describe forces and torque acting on the center of mass.

The rotor is modeled as rigid body, whereas the stator is split into a rigid lower part and a flexible upper part, because the moment of resistance of the lower part is much higher due to the increasing diameter. The mass of the AMB and the motor are integrated into the rigid part of the stator to reduce the complexity of the model. The upper PMB is modeled as point mass. As flexible beam model a Ritz approach based on a cubic function $u(z) = a_0 + a_1 z + a_2 z^2 + a_3 z^3$ is used to describe the occuring displacements of the flexible stator part. Figure 7 shows the comparison of the exact and approximated first and second Eigenmodes of a semibeam. As can be seen, the cubic approach gives a very good representation of the first bending mode which is the most important for the investigated system.

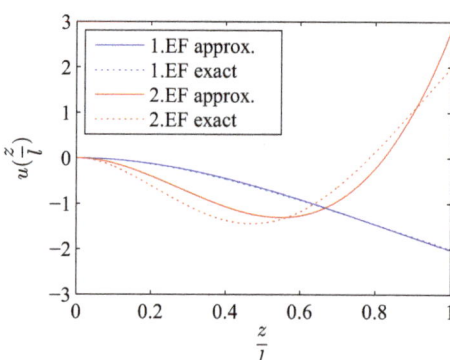

Figure 7. Comparison of the exact and approximated Eigenmodes of a semibeam bending.

For modelling a linear spring elements the potential energy is used. In the system such linear springs can be used as representation for

- the passive magnetic ring bearings,
- the stiffness of the damping elements,
- the (negative) stiffness of the motor,
- and the (negative) stiffness of the active magnetic bearing in radial direction.

The potential energy of a linear spring is determined by

$$V_F = \frac{1}{2}\Delta x(q) \, K \, \Delta x(q), \tag{12}$$

where $\Delta x(q)$ describes the relative extension vector of the spring from the force-free position as a function of the generalized coordinates q and K represents the tensor of spring constants. The same approach can also be used to describe the influence of elastic elements to their attachments. The speed-proportional potential energy of a damper is given by

$$R_D = \frac{1}{2}\Delta \dot{x}(q,\dot{q}) \, D \, \Delta \dot{x}(q,\dot{q}). \tag{13}$$

D describes the tensor of damping constants and $\Delta \dot{x}(q,\dot{q})$ the relative velocity vector of the damper dependent on q and $\dot{q}$.

The equation of motion obtained from the projection equation generally shows the structure

$$M(q)\ddot{q} + g(q,\dot{q}) - Q(q,\dot{q}) = 0. \tag{14}$$

and is of non-linear character. $M(q)$ is the skew symmetric mass matrix. $g(q,\dot{q})$ contains the remaining terms of the acceleration like centrifugal- and coriolis forces. In $Q(q,\dot{q})$ the applied spring-, damper- and weightforces as well as the actuating forces and moments take place. If only small deflections from the rest position $q = q_0 + \Delta q$ are assumed, Equation (14) can be developed by a TAYLOR series. A development up to terms of first order is sufficient and only the linear differential equation for small deflections remains

$$M_0 \Delta \ddot{q} + P_0 \Delta \dot{q} + Q_0 \Delta q = 0 \tag{15}$$

with

$$M_0 = M(q_0), \quad P_0 = \left.\frac{\partial(g-Q)}{\partial \dot{q}}\right|_{q_0}, \quad Q_0 = \left.\frac{\partial M \ddot{q}}{\partial q}\right|_{q_0} + \left.\frac{\partial(g-Q)}{\partial q}\right|_{q_0}. \tag{16}$$

The matrices P_0 and Q_0 can be split in their symmetric and antisymmetric parts whereby the linearized equation of motion is as follows:

$$M_0 \Delta \ddot{q} + (G+D)\Delta \dot{q} + (N+K)\Delta q = 0 \tag{17}$$

- G: gyroscopic matrix
- D: damping matrix
- N: matrix of non-conservative forces
- K: matrix of conservative forces (potential forces)

Due to the symmetry of the bearing arrangement, it is possible to halve the system order by introducing complex coordinates. The translations are defined by

$$\underline{r}_c = x + jy \tag{18}$$

and the tilting by

$$\underline{\varphi}_c = \alpha - j\beta. \tag{19}$$

The transformation $z_r = T_c \Delta q$ leads to the system valid in the complex coordinates z:

$$M_c \ddot{z}_r + (G_c + D_c)\dot{z}_r + (N_c + K_c)z_r = 0. \tag{20}$$

The system matrices are transformed according

$$A_c = T_c \cdot A \cdot \text{Re}\{T_c^T\}, \tag{21}$$

whereby A stands for a general matrix.

To include the viscoelastic behavior the dynamic system model with z_r states has to be extended with the inner states z_i of the material model

$$z = \begin{bmatrix} z_r & | & y_i & v_i \end{bmatrix}^T = \begin{bmatrix} z_r & | & z_i \end{bmatrix}^T. \tag{22}$$

There y_i are the translational and v_i the rotational inner states. The first order differenzial system is then given by

$$\begin{bmatrix} E & 0 & 0 \\ 0 & M_c & 0 \\ 0 & 0 & D_i \end{bmatrix} \begin{bmatrix} \dot{z}_r \\ \dot{z}_r^p \\ \dot{z}_i \end{bmatrix} + \begin{bmatrix} 0 & -E & 0 \\ K_c & D_c + G_c & K_{ic} \\ K_{ci} & 0 & K_i \end{bmatrix} \begin{bmatrix} z_r \\ z_r^p \\ z_i \end{bmatrix} = 0 \tag{23}$$

with $\dot{z}_r = z_r^p$. This transformation in a first order system is not trivial, because the inner states are massless and thus the mass matrix M is not invertible. As the inner states are only present in the first derivative only the general states z_r have to be transformed. The system of the inner states can be included in the last row and column of (23).

With the derived system the deflections of the system as well as the eigenfrequencies can be calculated. For the eigenmodes the linear differential equation

$$\dot{x} = Ax \tag{24}$$

can be solved using an exponential approach

$$x = \hat{x} e^{\lambda t} \tag{25}$$

leading to the characteristic system

$$[\lambda E - A]\, \hat{x} = 0. \tag{26}$$

This system has a non-trivial solution when

$$\det(\lambda E - A) = 0. \tag{27}$$

The eigenvalues resulting from the solution of the polynomials have thereby the shape

$$\lambda_i = \delta_i + j\omega_i. \tag{28}$$

δ_i describes the damping constant and ω_i the frequency of the corresponding eigenmode. The resulting motions of the system in its eigenmodes are depicted in Figure 8.

The green points represent the PMB's and the viscoelastic damping connections. The red cylinder is thereby showing the rotor and the blue shaft indicates the flexible stator beam connected to the grey rigid part of the stator. As it is shown in the four modes the stator is tilting as a entity, where as the rotor shows a tilting around the lower PMB in the first mode and around the upper PMB in the second mode. In the third rigid mode the rotor points a translational movement. The fourth mode indicates the semibeam bending of the flexible stator beam.

Figure 9 shows the effect of the flexible stator shaft of a designated system for different shaft materials. In the top PMB the deflections in the first resonance at about 40 Hz are increasing with softer material. Furthermore, the bending frequency, which is present in the range of 400–500 Hz is situated in the operation range. The stator shaft material has a huge influence on the relative rotor deflections in the first bending mode. Due to the required space limitation for the AMB, the motor and the position sensor system, the minimum stator shaft length is given with 116 mm as used in Figure 9. As the deflection, due to the bending, cannot be influenced by the stiffness of the PMBs and the stiffness and positioning of the damping elements, the only suitable material for the shaft, that meets the requirements of a maximum deflection in the PMB planes of 500 µm, is steel.

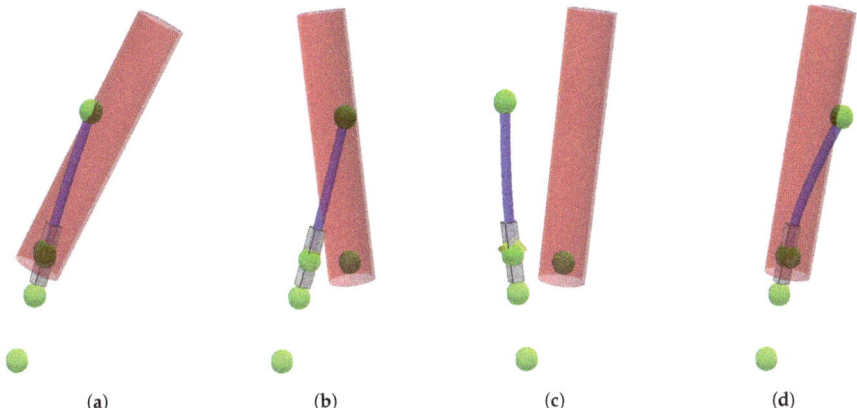

Figure 8. Schematic representation of the system motions in the three rigid body modes (**a**–**c**) and the first flexible mode of the stator beam (**d**).

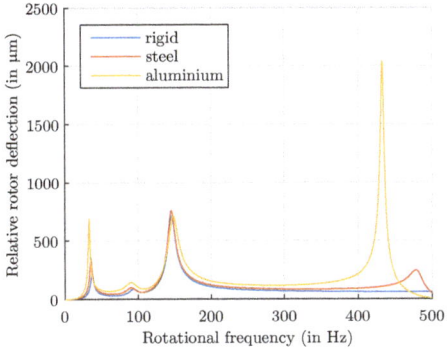

Figure 9. Relative rotor deflection at the upper PMB of not optimized systems with different shaft materials.

It is our goal to design a system that does not exceed a maximum deflection of 500 µm in both PMBs to prevent contact with the fault bearings in case of a process-unbalance of 20 gmm. To achieve this the stiffness and the position of the PMBs and the damping elements are varied for optimization.

5. Optimization

In the first optimization a given system setup was used and only the position and dimensions of the viscoelastic damping elements were optimized by minimizing the relative rotor deflections to the stator in the PMB planes. For optimization the software tool "SymSpace", developed at the Linz Center of Mechatronics (LCM), was used. It uses a genetic algorithm described in detail in [28]. In Table 1 the main parameters of the system setup are given. The mass of the rotor includes the additional mass which is mounted on the rotor. The variable parameters of the optimization with their referring range limits are shown in Table 2.

Table 1. Main parameters of the investigated system.

Description	Variable	Value	Unit
Rotor			
Mass of rotor	m_r	474	g
Polar moment of inertia	J_{rp}	1.1×10^{-4}	kg m^2
Diametrical moment of inertia	J_{rd}	4.74×10^{-3}	kg m^2
Stator			
Mass of stator	m_s	373	g
Diametrical moment of inertia	J_{rd}	8.2×10^{-4}	kg m^2
PMB's			
Radial stiffness of upper PMB	$k_{r,upper}$	25.7×10^3	N/m
Radial stiffness of lower PMB	$k_{r,lower}$	38.1×10^3	N/m
Magnetic air gap of PMB's	δ_m	0.8	mm
Geometric air gap of PMB's	δ_g	0.5	mm
AMB			
Radial stiffness of AMB	$k_{r,AMB}$	-5×10^3	N/m
Motor			
Radial stiffness of motor	$k_{r,motor}$	-6×10^3	N/m
Damping elements			
Outer damping elements diameter	$d_{o,DE}$	36	mm
Inner damping elements diameter	$d_{i,DE}$	23	mm

Table 2. Variables for optimization.

Description	Variable	Min.	Max.	Unit
Height of upper element	$h_{u,upperDE}$	5	20	mm
Height of lower element	$h_{l,upperDE}$	5	20	mm
Position of upper element to the center of mass of the stator	$l_{u,upperDE}$	−30	−50	mm
Position of lower element to the center of mass of the stator	$l_{l,upperDE}$	−50	−100	mm
Material of upper element		Shore A40, Shore A20		
Material of lower element		Shore A40, Shore A20		

For excitation, two different unbalances with an unbalance value of 20 gmm are assumed. They are located at the top and the bottom of the rotors linked part. Finally, it turned out that the employment of two different damping materials leads to the best results. For the bottom viscoelastic element the material with a hardness of Shore A40, which was measured, is preferable. For the upper damping viscoelastic element a softer material with a hardness of Shore A20 shows the best results. However, for this material only the manufacturer data was available leading to an uncertainty since the material parameters play a major role for the rotor dynamic behavior. In Figure 10 the Pareto front of the optimization is shown.

It can be seen that the relative rotor deflections can be minimized to a value of 160 µm for both unbalance positions, which is much smaller than the PMB air gap of 500 µm. This proves the high potential of elastomer materials. The optimized values for the damping elements are summarized in Table 3.

Thereby, the lower harder element has the positive effect that it will carry the axial pre-load generated from the system.

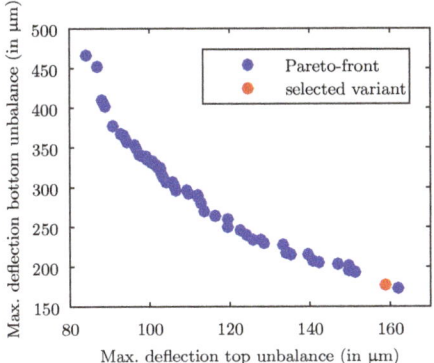

Figure 10. Pareto-front of the optimization referring the maximum deflections arising due to the specified unbalances of 20 gmm.

Table 3. Optimized Variables.

Variable	Value	Chosen	Unit
$h_{o,upperDE}$	9.6	10	mm
$h_{l,upperDE}$	5.01	5	mm
$l_{o,upperDE}$	−31.6	−32	mm
$l_{o,upperDE}$	−97.2	−97	mm
Material of upper element		Shore A20	
Material of damping element		Shore A40	

6. Verification of the System Model

To verify the system model a prototype was built with the optimized damping elements. To ensure a comparable initial situation to the simulation the rotor was initially balanced without any linked part. As the unbalance of the process mass is not predictable, the verification is done with a system without any process mass. Afterwards, two defined unbalances given in Table 4 were placed at two specified positions also used for the balancing of the system. As it is not easily possible to measure the relative deflections directly, only the absolute rotor movement is compared. The measurements are conducted using laser triangulation sensors with a sample time up to 100 kHz and a resolution of 25 nm. Hence, also high rotational speeds can be quantified with high resolution.

Table 4. Verification unbalances.

	Position	Value	Unit
Unbalance1	upper balance plane	7.6	g mm
Unbalance2	lower balance plane	5.6	g mm

To eliminate the effects of the remaining rotor unbalance, the deflections of the balanced rotor were subtracted from the deflections with the defined unbalances. It should be noted that this subtraction has to be performed properly considering the phases of the measured data and is only permitted assuming a linear behavior of the system, which is already assumed in the modeling.

In Figure 11 the measured absolute rotor deflections in the PMB planes (see Figure 1) over to the rotational rotor frequency is shown. It can be noticed that the highest displacement is occuring in the lower PMB due to the unbalance in the upper balance plane. The difference to the low optimized deflections shown in Figure 10 is caused by the lower rotor mass of the verification system as it is measured without the additional mass, which has an negative influence on the system adjustment as shown in [20].

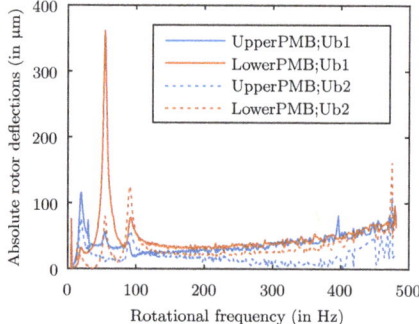

Figure 11. Measured deflections of the system according to applied unbalances as given in Table 4.

Thus, the measured rotor displacements need to be compared with an adapted simulation. For this the rotor has to be modified to characterize the one in the measured system without the process mass. Without the process mass the rotor features the data given in Table 5.

Table 5. Parameters of the varified rotor.

Description	Variable	Value	Unit
Mass of Rotor	m_r	365	g
Polar moment of inertia	J_{rp}	9.32×10^{-5}	kg m^2
Diametral moment of inertia	J_{rd}	2.46×10^{-3}	kg m^2

Comparing the adapted simulation with the measurement, as shown in Figure 12, revealed that the rigid body modes of the real system are situated at lower frequencies. That indicates a system featuring a lower overall stiffness. To identify the source of the difference first the stiffness of the PMBs are investigated. In a separated measurement of the PMBs it turned out, that the real stiffness is 20% lower than calculated. This is due to a lower PM remanent flux density, which was also confirmed with a dipole measurement of the PM rings.

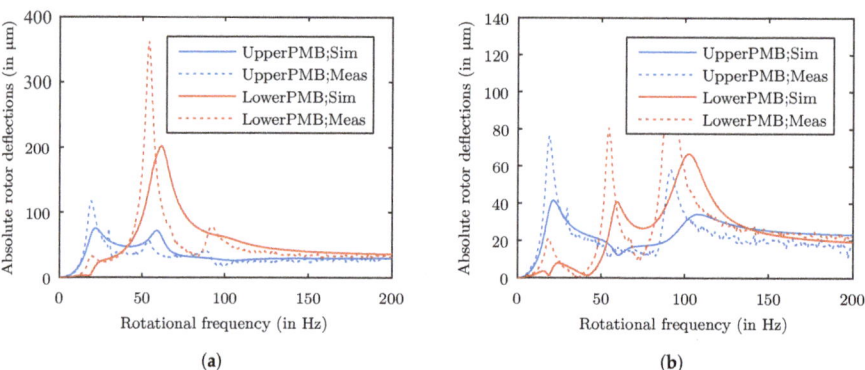

Figure 12. Comparison of the measured data with the simulated deflections with "Unbalance1" (**a**) and "Unbalance2" (**b**) specified in Table 4.

Furthermore, the difference in the Eigenfrequencies indicate that the material of the upper damping element offers a lower stiffness, either. To verify this assumption measurements of the used material with a hardness of Shore A20 at a temperature of 20 °C and 25 °C were performed and compared to the related stiffness and damping factor characteristics provided by the manufacturer.

Figures 13 and 14 show that the real stiffness and damping behavior is smaller than expected. The measured stiffness values is only be reflected when setting the manufacturer data to a 5 °C higher temperature. Regarding the loss factor of the material even about 12 °C temperature rise has to be performed. That result shows again the necessity for proper material data, since the damping directly affects the occurring deflections of the rotor. The large deviation is also due to the measuring method used by the manufacturer, which uses a prestressing of the material in comparison to the self-used measurement process without prestressing.

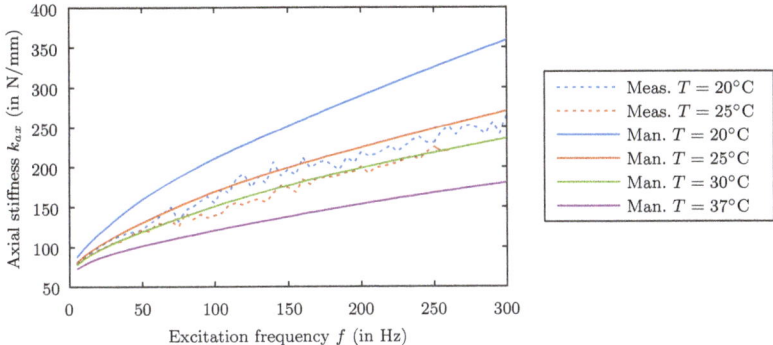

Figure 13. Comparison of measured probe stiffness with the manufacturer data.

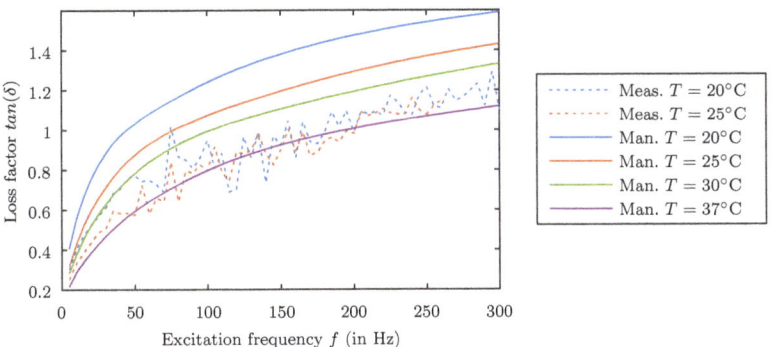

Figure 14. Comparison of measured loss factor with the manufacturer data.

In Figure 15 the comparison of the adjusted simulation is presented. The adopted temperature was set to a 12 °C higher temperature as the loss factor is the most important parameter considering the deflection amplitudes. The frequency range is limited to 200 Hz since the bending of the stator shaft is hardly apparent. The deflections show a very good agreement with the measurements.

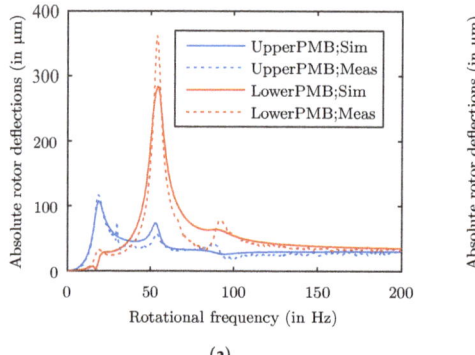 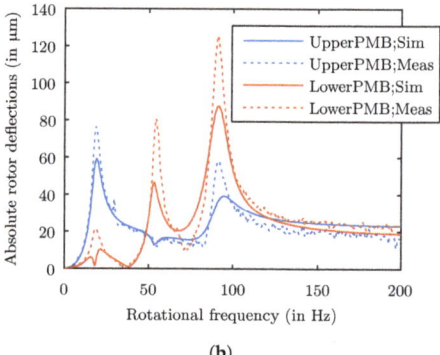

Figure 15. Comparison of the measured data with the simulated deflections with "Unbalance1" (**a**) and "Unbalance2" (**b**) specified in Table 4 considering the lower PMB stiffnesses and a adopted temperature of 37 °C for the upper damping element.

7. Discussion

Bringing magnetic bearing technology to a broader field of application requires the reduction of complexity and costs. The idea to use passive magnetic bearings, to achieve this goal, seems promising. So in the best case only one degree of freedom has to be controlled actively what saves costs for required actors, sensors and power electronics. Moreover, passive permanent magnetic bearings feature high reliability in case of power loss and failures. Disadvantageous proves the fact, that these type of bearings require a precise design regarding their field of application. Based on the nearly undamped system behavior, unbalances play a critical role. Different to active magnetic bearings no damping can be induced and adjusted by control, which is one of the main reasons for the limited use in commercial products. However, there are many ways to induce damping into such systems. To keep the cost advantage of passive magnetic bearings, a solution for damping must be found which only causes low additional costs. Using viscoelastic damping elements often failed in the past due to the very complex modeling and the hardly available and reliable manufacturers data. Even if available, the data is strongly influenced by the measurement process. As shown in this work, useful results can only be obtained when the exact characteristics of the damping material is well known. Unfortunately its determination is very time-consuming and an expensive task.

Moreover, a proper model of the rotordynamic behavior has to be derived to evaluate the occurring deflections of the system parts. Thereby, the most important deflections, which has to be considered, are the relative deflections in the PMB's. Since the air gap between the outer rings and the inner rings is limited, it has to be ensured, that any contact is prohibited. Thus, the generation of a proper all-over system model is complex. As there are numerous parameters influencing the system performance, the demand for an optimization is obvious. As is shown, the right configuration leads to very good results according the stabilizability of unbalances. However, only the knowledge of the exact system parameters guarantee a proper prediction of the dependent variables.

Author Contributions: Conceptualization, J.P.; validation, J.P.; formal analysis, J.P.; investigation, J.P.; writing—original draft preparation, J.P.; measurements, J.P.; writing—review and editing, G.J.; supervision, G.J.; funding acquisition, W.A.

Funding: This research received no external funding.

Acknowledgments: The presented research work has been supported by the Linz Center of Mechatronics GmbH (LCM), which is part of the COMET/K2 program of the Federal Ministry of Transport, Innovation and Technology and the Federal Ministry of Economics and Labor of Austria. The authors would like to thank the Austrian and Upper Austrian Government for their support.

Conflicts of Interest: The authors declare no conflict of interest.

References

1. Kailasan, A.; Dimond, T.; Allaire, P.; Sheffler, D. Design and Analysis of a Unique Energy Storage Flywheel System—An Integrated Flywheel, Motor/Generator, and Magnetic Bearing Configuration. *J. Eng. Gas Turbines Power* **2015**, *137*, 042505. [CrossRef]
2. Circosta, S.; Bonfitto, A.; Lusty, C.; Keogh, P.; Amati, N.; Tonoli, A. Analysis of a Shaftless Semi-Hard Magnetic Material Flywheel on Radial Hysteresis Self-Bearing Drives. *Actuators* **2018**, *7*, 87. [CrossRef]
3. Jungmayr, G.; Amrhein, W.; Grabner, H. Minimization of forced vibrations of a magnetically levitated shaft due to magnetization toleranzes. In Proceedings of the 12th International Symposium on Transport Phenomena and Dynamic of Rotating Machinery, Honolulu, HI, USA, 17–22 February 2008.
4. Delamare, J.; Yonnet, J.P.; Rulliere, E. A Compact Magnetic Suspension With Only One Axis Control. *IEEE Trans. Magn.* **1994**, *30*, 4746–4748. [CrossRef]
5. Lodge, W.S. *The Passive Characteristics of Repulsion Magnetic Bearing*; Techreport 76170; Royal Aircraft Establishment: Farnborough, UK, 1976.
6. Yonnet, J.P. Permanent Magnet Bearings and Couplings. *IEEE Trans. Magn.* **1981**, *17*, 1169–1173. [CrossRef]
7. Yonnet, J.P. Etude des Paliers Magnetiques Passives. Ph.D. Thesis, Universite de Grenoble, Saint-Martin-d'Hères, France, 1980.
8. Marinescu, M. Analytische Berechnung und Modellvorstellungen für Systeme mit Dauermagneten und Eisen. Ph.D. Thesis, TU Braunschweig, Braunschweig, Germany, 1979.
9. Lang, H. Fast calculation method of forces and stiffnesses of permanent-magnet bearing. In Proceedings of the 8th International Symposium on Magnetic Bearings, Mito, Japan, 26–28 August 2002.
10. Jungmayr, G. Der Magnetisch Gelagerte Lüfter. Ph.D. Thesis, Johannes Kepler University, Linz, Austria, 2008.
11. Chen, H.; Walter, T.; Wheeler, S.; Lee, N. A Passive Magnet Bearing System for Energy Storage Flywheels. In Proceedings of the 9th International Symposium on Magnetic Bearings, Lexington, Kentucky, 3–6 August 2004.
12. Filatov, A.V.; Hawkins, L.A.; Krishnan, V.; Law, B. Active Axial Electromagnetic Damper. In Proceedings of the 11th International Symposium on Magnetic Bearings, Nara, Japan, 26–29 August 2008.
13. Filatov, A.; Maslen, E.H. Passive Magnetic Bearing for Flywheel Energy Storage System. *IEEE Trans. Magn.* **2001**, *37*, 3913–3924, doi:10.1109/20.966127. [CrossRef]
14. Beams, J.W. Double Magnetic Suspension. *Rev. Sci. Instrum.* **1963**, *34*, 1071–1074. [CrossRef]
15. Moser, R.; Sandtner, J.; Bleuler, H. Optimization of Repulsive Passive Bearings. *IEEE Trans. Magn.* **2006**, *42*, 2038–2042. [CrossRef]
16. Earnshaw, S. On the Nature of Molecular Forces which regulate the Constitution of the Luminiferous Ether. *Trans. Camb. Philos. Soc.* **1842**, *7*, 97–112.
17. Braunbeck, W. Freischwebende Körper im elektrischen und magnetischem Feld. *Z. Phys.* **1939**, *112*, 753–763. [CrossRef]
18. Passenbrunner, J.; Jungmayr, G.; Panholzer, M.; Silber, S.; Amrhein, W. Simulation and Optimization of an Eddy Current Position Sensor. In Proceedings of the 11th IEEE International Conference on Power Electronics and Drive Systems (PEDS 2015), Sydney, Australia, 9–12 June 2015.
19. Passenbrunner, J.; Silber, S.; Amrhein, W. Investigation of a Digital Eddy Current Sensor. In Proceedings of the IEEE International Electric Machines and Drives Conference (IEMDC), Coeur d'Alène, ID, USA, 10–13 May 2015.
20. Marth, E.; Jungmayr, G.; Amrhein, W. Fundamental Considerations on Introducing Damping to Passively Magnetically Stabilized Rotor Systems. *Adv. Mech. Eng.* **2016**, *8*, doi:10.1177/1687814016682150. [CrossRef]
21. Bormann, A. Elastomerringe zur Schwingungsberuhigung in der Rotordynamik—Theorie, Messungen und optimierte Auslegung. Ph.D. Thesis, Technische Universität Berlin, Berlin, Germany, 2005.
22. Osswald, T. *Understanding Polymer Processing*; Hanser: München, Germany, 2010.
23. Bonfitto, A.; Tonoli, A.; Amati, N. Viscoelastic Dampers for Rotors: Modeling and Validation at Component and System Level. *Appl. Sci.* **2017**, *7*, 1181. [CrossRef]
24. Gutierrez-Lemini, D. *Engineering Viscoelasticity*; Springer: Berlin, Germany, 2014.
25. Ecker, R. Der Einfluss verstarkender Russe auf die viskoelastischen Eigenschaften von amorphen Elastomeren. *Kautsch. Gummi Kunstst.* **1968**, *21*, 304–317.
26. Ferry, J.D. *Viscoelastic Properties of Polymers*, 3rd ed.; John Wiley & Sons Ltd.: New York, NY, USA, 1970.

27. Bremer, H. *Elastic Multibody Dynamics*; Springer: Berlin, Germany, 2008.
28. Silber, S.; Koppelstätter, W.; Weidenholzer, G.; Bramerdorfer, G. MagOpt—Optimization Tool for Mechatronic Components. In Proceedings of the ISMB14, 14th International Symposium on Magnetic Bearings, Linz, Austria, 11–14 August 2014; pp. 243–246.

© 2019 by the authors. Licensee MDPI, Basel, Switzerland. This article is an open access article distributed under the terms and conditions of the Creative Commons Attribution (CC BY) license (http://creativecommons.org/licenses/by/4.0/).

Article

Performance Enhancement of a Magnetic System in a Ultra Compact 5-DOF-Controlled Self-Bearing Motor for a Rotary Pediatric Ventricular-Assist Device to Diminish Energy Input [†]

Masahiro Osa [1],*, Toru Masuzawa [1], Ryoga Orihara [1] and Eisuke Tatsumi [2]

[1] Domain of Mechanical Systems Engineering, Graduate School of Science, Ibaraki University, 4-12-1, Nakanarusawa, Ibaraki, Hitachi 319-1222, Japan; toru.masuzawa.5250@vc.ibaraki.ac.jp (T.M.); orihara.r9@gmail.com (R.O.)
[2] Department of Artificial Organs, National Cerebral and Cardiovascular Center Research Institute, 5-7-1, Fujishiro-dai, Suita, Osaka 565-8565, Japan; tatsumi@ncvc.go.jp
* Correspondence: masahiro.osa.630@vc.ibaraki.ac.jp; Tel.: +81-294-38-5030
[†] This paper is an extended version of our paper published in: Osa, M.; Masuzawa, T.; Orihara, R.; Tatsumi, E. Magnetic suspension performance enhancement of ultra-compact 5-DOF-controlled self-bearing motor for rotary pediatric ventricular assist device. In Proceedings of the 16th International Symposium on Magnetic Bearings (ISMB16), Beijing, China, 13–17 August 2018.

Received: 15 February 2019; Accepted: 13 April 2019; Published: 15 April 2019

Abstract: Research interests of compact magnetically levitated motors have been strongly increased in development of durable and biocompatible mechanical circulatory support (MCS) devices for pediatric heart disease patients. In this study, an ultra-compact axial gap type self-bearing motor with 5-degrees of freedom (DOF) active control for use in pediatric MCS devices has been developed. The motor consists of two identical motor stators and a centrifugal levitated rotor. This paper investigated a design improvement of the magnetic circuit for the self-bearing motor undergoing development in order to diminish energy input by enhancing magnetic suspension and rotation performances. Geometries of the motor were refined based on numerical calculation and three-dimensional (3D) magnetic field analysis. The modified motor can achieve higher suspension force and torque characteristics than that of a previously developed prototype motor. Oscillation of the levitated rotor was significantly suppressed by 5-DOF control over rotating speeds up to 7000 rpm with lower energy input, indicating efficacy of the design refinement of the motor.

Keywords: axial gap; self-bearing motor; double stator structure; 5-degrees of freedom active control; design refinement; ventricular assist device; pediatric

1. Introduction

Mechanical circulatory support (MCS) is widely used for heart disease therapy. However, clinically available and implantable MCS devices are not applicable for pediatric patients, which have small body surface area (BSA < 0.7 mm^2) due to anatomical limitations [1,2]. Almost all pediatric patients have to rely on using EXCOR pediatric ventricular assist device (VAD), which is a extracorporeal pulsatile flow pump developed by Berlin Heart Inc. in Germany [3]. Pulsatile devices with diaphragm and valve configurations potentially limit a device's lifetime and have the risk of thrombosis. Currently, there have been increasing research interests in pediatric heart treatment with implantable rotary MCS devices specifically designed for pediatric circulatory support [4]. In 2010, MCS device development for pediatric patients was supported as a national project named PumpKIN (Pump for kids, infants and neonates) in the United States (US) [1,2,4]. A tiny rotary MCS device (Jarvik2015) for pediatric circulation is being developed by Jarvik Heart Inc. in the US [5–7]. However, the Jarvik device is now facing technical difficulties such as deterioration of mechanical durability, blood clotting and blood destruction, due to a mechanically contacting bearing to suspend a spinning rotor impeller. Hence, the development of next-generation MCS devices that can completely levitate a rotating impeller are in demand due to its high durability and better blood compatibility.

A magnetic suspension system is one of the strongest candidates to suspend the rotating impeller without mechanical contact. The biggest advantages of maglev technology are its high-speed motor drive, less heat generation due to no material wear, zero friction, less blood trauma, anti-thrombogenicity and increased mechanical reliability. In one of the earliest studies, 5-DOF-controlled maglev motors were developed [8,9]. The maglev system indicated high suspension stability, however, the device needed to be bigger due to the larger number of actuators: two radial magnetic bearings and two axial magnetic bearings. Miniaturization of the magnetic suspension system has a significant role in success of pediatric MCS device development. Reduction of actively controlled positions is a general strategy to miniaturize the magnetic suspension systems. For example, single DOF-controlled maglev systems, which are utilizing passive stability using magnetic coupling force or repulsive PM magnetic bearings to suspend the non-controlled DOF, were developed to achieve a simple structure and compact device size [10–16]. However, a larger number of passively stabilized axes potentially causes significant deterioration of suspension stability of the magnetic system. Hence, 2-degrees of freedom (DOF)-controlled radial maglev motors and 3-DOF-controlled double stator axial maglev motors have often been developed in blood pump applications [17–24]. These miniaturized maglev motors are successfully applied to implantable and extracorporeal MCS devices for adult patients, whereas further miniaturization of the maglev motors is required for use in rotary pediatric MCS devices. In extremely small magnetic suspension systems, there is a limitation of passive stabilization in multiple axes because the system cannot have sufficient capacity of passive stiffness and suspension force due to few spaces to have enough permanent magnet volume and turn number of control windings. Hence, a technical breakthrough is needed to achieve an ultra-compact magnetic system.

This study developed a compact pediatric MCS device with a novel self-bearing motor utilizing a 5-DOF-control concept that was newly developed in our laboratory, and the developed device demonstrated noncontact suspension and sufficient pump performance [25–28]. However, further improvement of magnetic suspension stability is necessary to achieve higher mechanical reliability and energy conservation system required for clinically applicable MCS devices. In this paper, design improvement of magnetic circuit for the 5-DOF-controlled self-bearing motor was investigated to enhance the magnetic suspension performance and energy efficiency by using theoretical calculation and three-dimensional (3D) finite element method (FEM) analysis. Static force and torque characteristics [28], dynamic suspension performance and energy consumption of the improved maglev motor with 5-DOF control are evaluated.

2. Materials and Methods

2.1. 5-DOF-Controlled Self-Bearing Motor for Pediatric Ventricular Assist Device (VAD)

2.1.1. Over View of Maglev Pediatric VAD with 5-DOF-Controlled Self-Bearing Motor

The 5-DOF-controlled self-bearing motor is driven as an axial gap type surface permanent magnet synchronous motor, which has 6-slot and 4-pole structure. Figure 1 shows a schematic of the self-bearing motor which is consists of two identical motor stators and a levitated impeller. The levitated impeller is axially suspended with the both stators. Integrated windings for suspension and rotation control are wound on each stator tooth. A motor torque and a suspension force are produced with double stator mechanism which can enhance motor torque and radial passive stability.

An axial position (z) and rotating speed (ω_z) are actively regulated with a 4-pole control magnetic field. Radial positions (x and y) and tilting angles (θ_x and θ_x) are actively regulated with a 2-pole control magnetic field. 5-DOF of impeller postures are independently regulated by overlapping the different control magnetic fields in the magnetic gap [27]. A developed centrifugal blood pump for pediatric patients can regulate flow rate from 0.5 to 2.5 L/min against head pressure of around 100 mmHg at rotating speeds of 4500–5500 rpm.

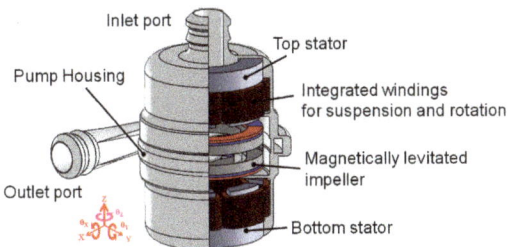

Figure 1. Structure of pediatric ventricular assist device with the axial gap type double stator 5-degrees of freedom (5-DOF)-controlled self-bearing motor.

2.1.2. Characterization of Suspension Force and Torque

The motor produces axial suspension force and rotating torque with a single rotating magnetic field based on vector control algorithm. An axial position (z) of the levitated impeller is actively regulated by field strengthening and field weakening as shown in Figure 2. A rotating speed (ω_z) of the rotor is controlled by conventional q-axis current regulation. The axial suspension force and the rotating torque are linearly produced with d-axis current id and q-axis current iq.

$$F_z = k_{Fz}(i_d - i'_d) \tag{1}$$

$$T_{\theta z} = k_{T\theta z}(i_q + i'_q) \tag{2}$$

Inclination angles (θ_x and θ_y) and radial positions (x and y) of the levitated rotor can be controlled with p ± 2 pole algorithm. The control magnetic field can simultaneously produce an inclination torque and a radial suspension force. Inclination torque around the y-axis and the radial suspension force in x direction are produced with the double stator mechanism as shown in Figure 3. The magnitude and the direction of the inclination torque and the radial suspension force can be linearly regulated with respect to excitation current supplied to the top stator and the bottom stator as following equations.

$$T_{\theta x/y} = k_{T\theta x/y}(i_{top} + i_{bottom}) \tag{3}$$

$$F_{x/y} = k_{Fx/y}(i_{top} - i_{bottom}) \tag{4}$$

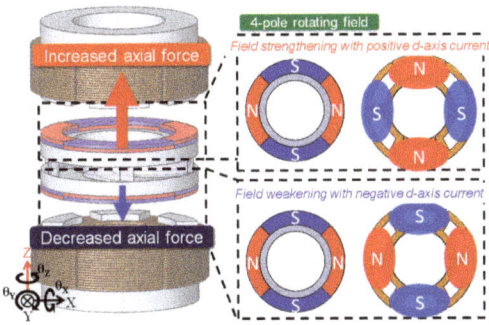

Figure 2. Axial position control by utilizing field strengthening and field weakening.

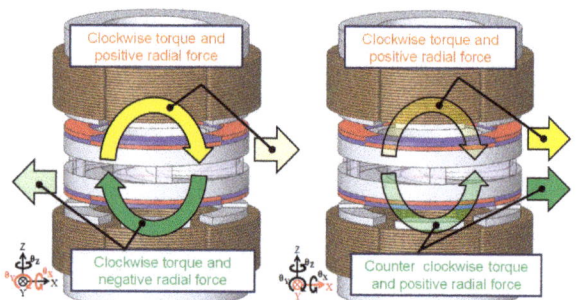

Figure 3. Inclination and radial position control with p ±2 pole rotating magnetic field.

2.2. Suspension Force and Torque Enhancement with Modified Magnetic Circuit of the Maglev Motor

2.2.1. Design Strategy of Suspension Performance Enhancement

The motor uses magnetic flux density produced by the permanent magnet as a main flux density to produce the suspension force and the rotating torque. Enhancement of the permanent magnet flux contributes to higher suspension force production. However, in miniaturized motor, there is difficulty to have sufficiently large cross-sectional area of the magnetic flux path, and it has possibility of deterioration of magnetic suspension force due to the magnetic saturation. Furthermore, excessively increased negative stiffness due to the high magnetic intensity potentially deteriorate controllability of the magnetic system. Hence, well design of the magnetic circuit which can keep a good balance between magnetic flux density produced by the permanent magnet and electromagnet is required to achieve sufficient magnetic suspension stability. In this study, design goal is to enhance the suspension force produced by the electromagnet without change of non-excited axial attractive force for avoiding instability caused due to the axial negative stiffness.

2.2.2. Design Refinement of Magnetic Circuit for the 5-DOF-Controlled Self-Bearing Motor to Enhance the Suspension Force, the Motor Torque and Reduce the Energy Input

Design improvement of a magnetic circuit for the 5-DOF-controlled self-bearing motor was performed based on following design strategy to enhance the magnetic suspension performance. (1) Keeping device size such as the outer diameter of 22 mm, the total height of 33 mm and the total volume of the previously developed prototype motor. (2) Maintaining the axial negative stiffness k_z within ±10% of that produced by the previously developed prototype motor. (3) Maximizing the force coefficient in the axial direction k_i defined as a slope of the suspension force to excitation current.

Geometries representatively characterizing the magnetic circuit of the self-bearing motor: pole height l_p, pole cross sectional area A_p, magnetic gap length l_g and permanent magnet thickness l_m, were numerically designed with theoretical calculation and fixed by using 3-D FEM magnetic field analysis as shown in Figure 4. Height and cross-sectional area of the stator pole were determined as 9.3 mm and 17.0 mm^2 based on the theoretical calculation. These geometries can increase in a turn numbers of coils and effectively maximize the force coefficient with slight change of the negative stiffness and the non-excited force. Parametric study in the magnetic gap length and the PM thickness were then performed. Variable parameters of the magnetic gap length of 1.3–1.7 mm and the PM thickness of 0.8–1.2 mm were chosen considering fabrication. Each combination of the magnetic gap length and PM thickness were simulated, and the non-excited axial negative stiffness k_z and the force coefficient k_i were estimated. A suspension index is defined as ratio of the force coefficient to the negative stiffness, which indicates rotor displacement with respect to excitation current of 1 A. The geometry which can achieve the biggest suspension index and satisfy the above design strategy was chosen as optimal design of the motor, which achieves well suspension performance with lower energy input.

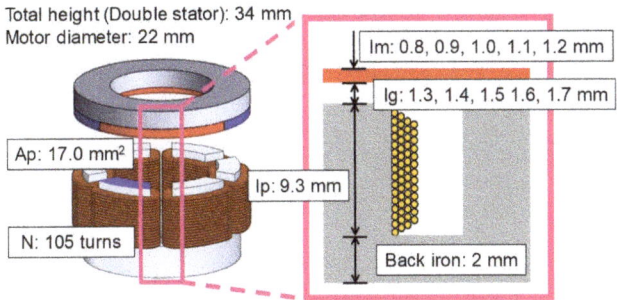

Figure 4. Variable geometries of magnetic circuit for the self-bearing motor in design improvement using 3-D finite element method (FEM) magnetic field simulation.

The numerically estimated force coefficient and non-excited attractive force of the self-bearing motor with different geometries in the magnetic gap length and the PM thickness are shown in Figure 5. The suspension index in each motor geometry is listed in Table 1. Red and yellow colored cells indicate satisfying design requirements in the negative stiffness and the force coefficient. The optimal geometry in the FEM simulation to maximize the force coefficient (k_i < 1.2 N/A) and maintain the axial negative stiffness (15.2 N/mm < k_z < 18.6 N/mm) is the shortest magnetic gap length of 1.3 mm and the thinnest PM of 0.8 mm as shown in red colored cell in Table 1. The force coefficient and the negative stiffness of the optimally designed motor are 2.0 N/A and 17.1 N/mm. Deterioration of the magnetic flux density with reduction of the PM thickness can be compensated by reducing the magnetic gap length. The shorter magnetic gap length and the thinner PM thickness can reduce magnetic resistance for the electromagnet and effectively enhance the magnetic suspension force production with excitation current. The geometries of the previously developed prototype motor and the improved motor with the final design are summarized in Table 2.

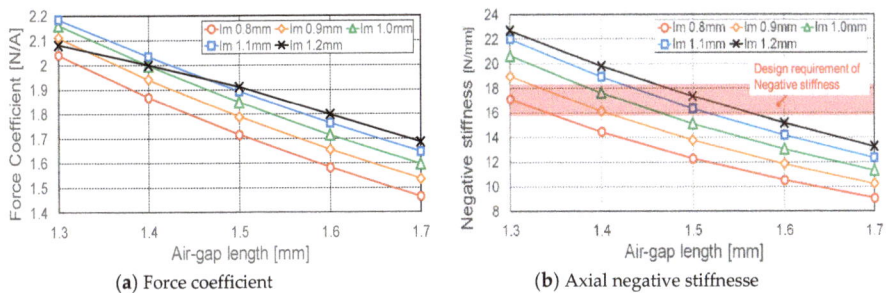

(a) Force coefficient   (b) Axial negative stiffnesse

Figure 5. Force coefficient and negative stiffness of the motor with different magnetic gap length and permanent magnet thickness.

Table 1. Estimated results of suspension index with different permanent magnet thickness and magnetic gap length.

Magnetic Gap Length [mm]	Permanent Magnet Thickness [mm]				
	0.8	0.9	1.0	1.1	1.2
1.3	0.119	0.111	0.104	0.099	0.092
1.4	0.130	0.120	0.114	0.108	0.101
1.5	0.139	0.130	0.122	0.116	0.110
1.6	0.152	0.139	0.132	0.125	0.119
1.7	0.161	0.149	0.141	0.133	0.127

Table 2. Motor geometries of the prototype motor and the improved motor.

Geometric Parameters	Prototype Motor	Improved Motor
Stator/rotor inner diameter [mm]	16	16
Stator/rotor outer diameter [mm]	22	22
Single stator height (total) [mm]	11.3	11.3
Stator teeth height [mm]	7.3	9.3
Stator back iron height [mm]	4	2
Magnetic gap length [mm]	1.5	1.3
Rotor back iron thickness [mm]	2	2
Permanent magnet thickness [mm]	1	0.8
Number of coil turns for each teeth	66	105

2.3. Developed System of 5-DOF-Controlled Maglev Motor with Modified Magnetic Circuit

2.3.1. Fabrication of 5-DOF-Controlled Maglev Motor

A 5-DOF-controlled self-bearing motor for pediatric VAD shown in Figure 6 was developed referring to motor geometries determined by using 3D FEM magnetic field analysis. The outer diameter and the total height are 22 mm and 33 mm. The length of magnetic gap of the developed motor is set to 1.3 mm. The material used for magnetic core of the motor stator and the rotor back iron is soft magnetic iron (SUY-1). The permanent magnets of 0.8 mm thickness are made of Nd-Fe-B, that has coercivity and residual flux density of 907 kA/m and 1.36 T, respectively. Concentrated cupper windings of 105 turns are wound on each stator tooth. Pump clearance between the pump casing and levitated rotor in the axial and radial direction are 0.3 mm and 0.5 mm.

Figure 6. Developed 5-DOF-controlled maglev motor with modified magnetic circuit. Pump casing with sensor holder, motor stator and rotor with permanent magnets.

2.3.2. Control System for Magnetic Levitation and Rotation with Digital PID Controllers

Digital PID controllers are implemented on a microprocessor board MicrolabBox (dSPACE GmbH, Paderborn Germany) with MATLAB/Simlink for 5-DOF active control. Figure 7 shows a schematic diagram of a 5-DOF control system. An axial position and inclination angles around the x and y axes of the levitated rotor are measured by three eddy current sensors (PU-03A, Applied Electronics Corporation). Radial positions of the levitated rotor in x and y direction are measured with other two eddy current sensors. A rotating angle of the levitated rotor is determined by outputs of three Hall effect sensors (Asahi KASEI Corporation) with a sensitivity of a 30-degree electrical angle. The rotating speed is calculated by time derivative of the rotating angle. Required current to produce the control magnetic flux density integrating three phase two-pole field and three-phase four-pole field synchronized with rotating PM field is calculated with PID controllers and is independently supplied to each coil by power amplifier (PA12A, Apex Microtechnology Corporation). Sampling and control frequency is 10 kHz. Control gains of the digital PID controllers for magnetic suspension and rotation were determined based on the previously measured motor suspension force and torque characteristics, and then, manually tuned in dynamic performance evaluation.

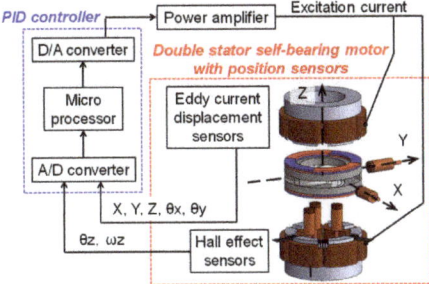

Figure 7. Schematic diagram of control system for 5-DOF-controlled maglev motor with position sensor, PID controller implemented in the microprocessor and power amplifier.

A block diagram for axial position and rotation control, inclination angle and radial position control are shown in Figures 8 and 9, respectively. Positive and negative d-axis current are determined by a PID feedback loop to produce an axial suspension force. q-axis current of both stators is regulated with a PI feedback loop for a conventional rotating speed control. Required current for inclination angle and radial position control are calculated by the other four PID feedback loops to determine amplitude and phase angle of two-pole rotating magnetic field produced by the top and bottom stators. PID gains for the position control and PI gains for the rotating speed control were set using a limit sensitivity method, and then manually tuned as shown in Table 3. Control gains of PID/PI controller of the previously developed prototype motor are also listed in Table 3.

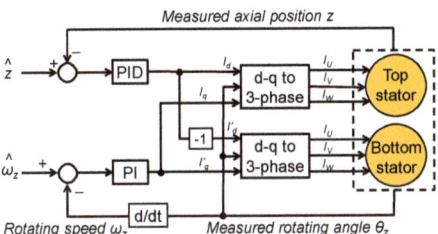

Figure 8. Block diagram of feedback loop for the axial position and the rotating speed regulation with d-q current regulation.

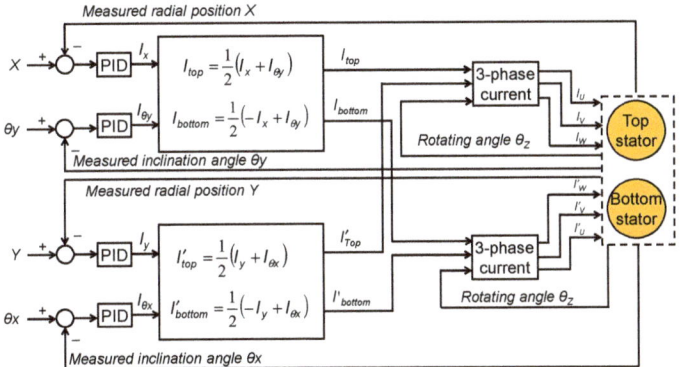

Figure 9. Block diagram of feedback loop for the inclination and the radial position control based on P +/- 2 pole algorithm.

Table 3. Control gains for impeller positioning and rotating speed regulation.

Controlled Axis		Axial Position	Radial Position	Inclination Angle	Rotating Speed
Prototype motor	P	13 [A/mm]	1.5 [A/mm]	3.5 [A/deg]	0.0005 [A/rpm]
	I	0.08 [A/mm sec]	0 [A/mm sec]	0.07 [A/deg sec]	0.0018 [A/rpm sec]
	D	0.01 [A sec/mm]	0.002 [A sec/mm]	0.0015 [A sec/deg]	-
Improved motor	P	10 [A/mm]	1.5 [A/mm]	1.5 [A/deg]	0.0005 [A/rpm]
	I	0.05 [A/mm sec]	0 [A/mm sec]	0.004 [A/deg sec]	0.0018 [A/rpm sec]
	D	0.007 [A sec/mm]	0.002 [A sec/mm]	0.001 [A sec/deg]	-

2.4. Magnetic Suspension Performance Evaluation of the Newly Developed Maglev Motor

2.4.1. Magnetic Flux Distribution Measurement and Static Magnetic Suspension Force and Torque Characteristics Measurement

Magnetic flux distribution produced by the rotor permanent magnets in the magnetic gap was measured without excitation. Four/two pole magnetic flux density produced by the electromagnet at excitation current of 1 A were then measured without the rotor permanent magnets. After that, static magnetic suspension force and torque characteristics: an axial negative stiffness k_z, a radial stiffness k_r, and suspension force of the designed motor was evaluated at excitation current of 1 A and magnetic gap length of 1.3 mm. The axial and radial suspension force were measured with load cell, and the inclination torque was calculated as a product of the measured force and rotor radius.

2.4.2. Dynamic Characteristics of Developed 5-DOF-controlled Maglev Motor

The rotor was magnetically levitated in water medium with 5-DOF control. The water flow was shut off by closed outlet port of the centrifugal pump to evaluate basic magnetic suspension characteristics by minimizing hydraulic force disturbance. Magnetic suspension performance with respect to increase in the rotating speed of the rotor was evaluated. The rotating speed was increased from 1000 rpm to 7000 rpm. Oscillation amplitude in axial direction and radial direction, maximum inclination angle around x and y axes, and power consumption of the motor during magnetic levitation and rotation were evaluated. The maximum oscillation amplitude was defined as half of the peak-to-peak value of rotor vibration.

3. Results

The measured magnetic flux density is shown in Figure 10. The magnetic flux density produced by the rotor permanent magnets did not significant difference between the prototype motor and the improved motor. In contrast, the magnetic flux density produced by the electromagnet of the improved motor significantly increased. The peak of the four/two pole magnetic flux density produced by the improved motor increased by 61% and 76% compare to the prototype motor.

Static suspension characteristics: stiffness, suspension force and torque, of the developed maglev motor which has the modified magnetic circuit and the previously developed maglev motor are shown together in Figure 11. The axial negative stiffness of the improved motor decreased by 17%, however, the radial stiffness was not significantly changed. The deterioration of the radial passive stability did not occur. The axial suspension force increased by 50 %, and the radial suspension force slightly decreased. Both the inclination torque and the rotating torque increased by 84% and 34%, respectively.

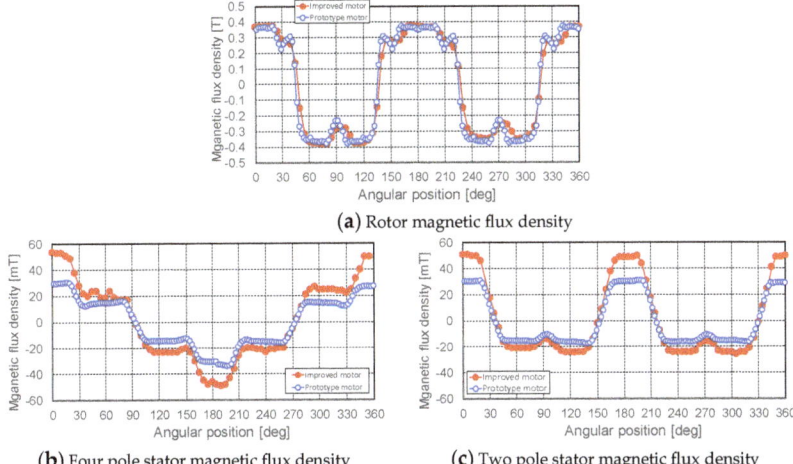

Figure 10. Magnetic flux density distribution in the magnetic gap. (**a**) Four pole magnetic flux density produced by the rotor magnet. (**b**) Four pole magnetic flux density for the axial position and the rotation control. (**c**) Two pole magnetic flux density for the radial position and the inclination control.

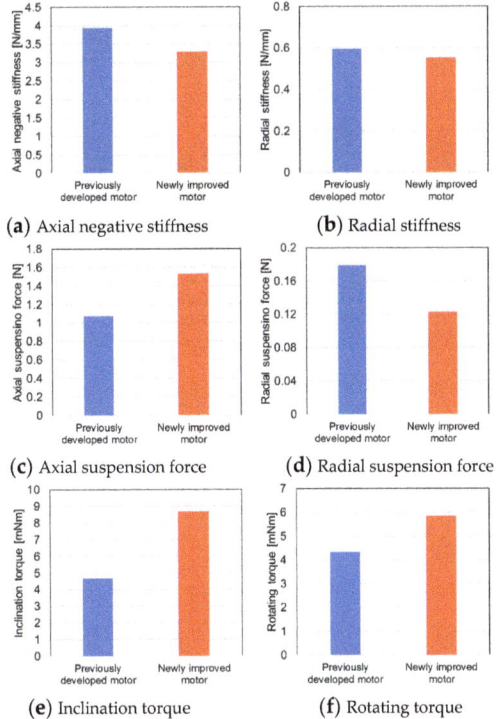

Figure 11. Static suspension characteristics of the developed maglev motor. (**a**,**b**) Axial and radial stiffness. (**c**,**d**) Magnetic suspension force in axial and radial direction at excitation current of 1 A. (**e**,**f**) Torque characteristics at excitation current of 1A.

The improved motor successfully achieves non-contact levitation and rotation up to the rotating speed of 7000 rpm. The maximum axial and radial oscillation amplitude and the maximum inclination angle around x and y axes with respect to the increase in the rotating speed of the levitated rotor are shown in Figures 12–14. In the lower speed range of 1000–3000 rpm, the oscillation amplitude of the levitated rotor was slightly increased in the prototype motor. In contrast, the oscillation amplitude in axial and radial direction, and the inclination angle of the improved motor were significantly suppressed around 20 µm, 100 µm and 0.4 degrees over every operational speed by enhancement of the suspension characteristics. The power consumption of the developed motor during magnetic levitation and rotation was shown in Figure 15, and it was in the range of 1–6 W at the rotating speeds of 1000–7000 rpm.

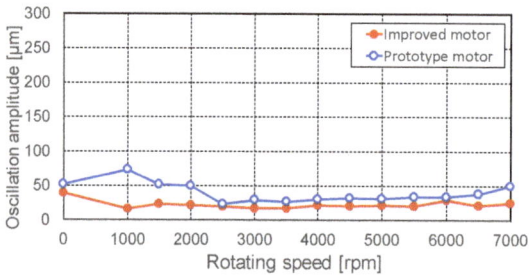

Figure 12. Maximum oscillation amplitude of the levitated rotor in axial direction with respect to increase in the rotating speed.

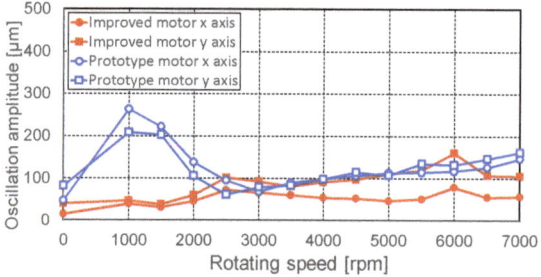

Figure 13. Maximum oscillation amplitude of the levitated rotor in radial direction with respect to increase in the rotating speed.

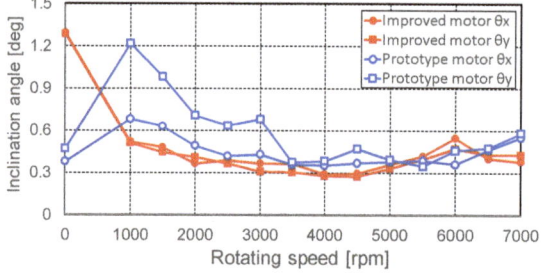

Figure 14. Maximum inclination angle of the levitated and rotated rotor around x and y axes with respect to increase in the rotating speed.

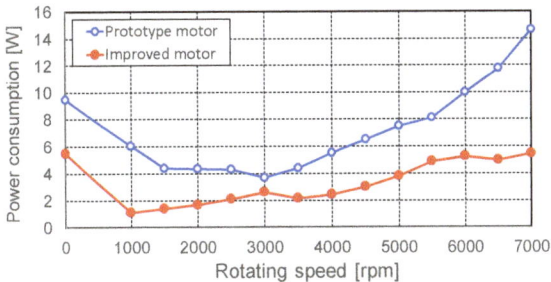

Figure 15. Power consumption of the developed maglev motor with respect to the increase in the rotating speed.

4. Discussion

Impeller suspension technique using magnetic suspension much contributes to enhance device durability and blood compatibility of the rotary MCS devices. In an ultra-compact maglev motor, optimization of the magnetic circuit for suspension system plays a significant role in the next generation rotary pediatric VADs development.

The lower negative stiffness of the magnetic system can be effective to reduce the suspension index, whereas the decreased stiffness will cause the deterioration of the passive stability in rotor radial direction and the motor torque. The negative stiffness was strategically adjusted to be maintained by keeping peak level of the magnetic flux density produced by the rotor permanent magnets in order to avoid the motor deterioration above mentioned in this study. Decrease in the pole surface area, magnetic gap length and permanent magnet thickness successfully played a significant role in enhancement of the magnetic flux density produced by the electromagnet due to increase in the turn numbers of coils and reduction of the magnetic resistance of the magnetic circuit. Magnetic saturation in the rotor iron could not be occurred because the magnetic flux density ratio of the improved motor and the prototype motor almost uniform in arbitrary angle in the air-gap.

The developed maglev motor successfully achieves the much higher suspension force coefficient k_i maintaining the negative stiffness k_z in the axial suspension characteristics. The suspension index of $k_i/k_z = 0.47$ mm/A is significantly higher than that of the previously developed motor ($k_i/k_z = 0.27$ mm/A). The axial negative stiffness was lower than estimated result. One of the causes of the above may have been deteriorated permanent magnet flux caused by reduced magnet volume due to coating thickness. Although the radial suspension characteristics slightly decreased, the deterioration of the total magnetic suspension performance will not occur because the magnitude of the radial suspension force is absolutely small. Even the radial suspension force produced by the newly developed motor is much effective to suppress the resonance and disturbance. The grossly increased inclination torque and the rotating torque due to the increase in the control magnetic flux density will contribute to achieve better suspension stability and low energy consumption.

The axial and radial oscillation amplitude and the inclination angle of the levitated rotor were small enough for rotary blood pump operation to prevent blood trauma. The improved motor demonstrated better magnetic suspension performance with the lower PID control gains than that of the previously developed prototype motor, that is indicating efficacy of the magnetic circuit design refinement to achieve higher mechanical reliability and lower energy input. First, resonance around 50 Hz, which is calculated from the measured radial stiffness and mass of the rotor implies that the resonance will not influence actual pump operation due to lower frequency than the operational frequency of the pediatric pump. The oscillation of the impeller was well suppressed, and resonance peak was not found at any rotating speed. Safeness against the resonance was experimentally verified. Frequency response measurement in all actively controlled axes should be required to investigate advanced dynamic characteristics as a next step.

The power consumption at the operating speed range (4000–5500 rpm) of the pediatric VAD was 2.4–4.9 W. The improved motor with modified magnetic circuit achieved more than 50% reduction of the power consumption compare with that of the previously developed motor. The decreased power consumption should be small enough for pediatric pump operation. The input power increased as increase in the rotating speed. This is due to increase in the copper loss caused by increased suspension current at higher speed rotation and the iron loss. Material change of the stator magnetic core will effectively reduce power consumption at higher rotating speed. In actual pump operation, required rotating torque increases to produce hydraulic output. In contrast, power consumption for suspension may become smaller due to reduced rotor oscillation by higher viscus damping of blood. The blood is filled in the pump cavity, however, magnetic properties of the blood do not affect to the magnetic performance of the self-bearing motor. Total energy input will be evaluated during circulation in future.

5. Conclusions

The ultra-compact 5-DOF-controlled self-bearing motor has been developed for pediatric MCS devices. Shortened magnetic gap and PM thickness effectively increased the control magnetic flux density due to the reduction of magnetic resistance maintaining PM magnetic flux density. In addition, increase in the turn number of the control windings almost double also played a significant role in enhancement of the control flux density production. As a result, the static magnetic suspension characteristics were successfully increased by up to 34–84% by refining the magnetic circuit of the motor. The dynamic magnetic suspension performance and further stable magnetic suspension with high speed rotation is successfully indicated due to the improved force and torque capacity with respect to the excitation current. Energy input was drastically reduced by less than half (1–5 W) because of the reduction of copper loss with low input current. The results indicate the efficacy of the magnetic circuit refinement of the proposed 5-DOF-controlled self-bearing motor. As a next step, dynamic suspension characteristics during pumping in actual circulation condition and pump performance of pediatric rotary VAD will be investigated.

Author Contributions: M.O. wrote the paper; T.M. and E.T. contributed in designing the experimental devices; M.O. performed 3D FEM analysis; M.O. and R.O. conducted the experiments; M.O. analyzed the numeric simulation and experimental results.

Funding: This work was supported by Japanese Society for the Promotion of Science (JSPS) KAKENHI Grant-in-Aid for Young Scientists (B) Grant Number 16K18036.

Conflicts of Interest: The authors declare no conflicts. The funders had no role in the design of the study; in the collection, analyses, or interpretation of data; in the writing of the manuscript, or in the decision to publish the results.

References

1. Baldwin, J.T.; Borovetz, H.S.; Duncan, B.W.; Gartner, M.J.; Jarvik, R.K.; Weiss, W.J.; Hoke, T.R. The National Heart, Lung, and Blood Institute Pediatric Circulatory Support. *Circulation* **2006**, *113*, 147–155. [CrossRef] [PubMed]
2. Baldwin, J.T.; Borovetz, H.S.; Duncan, B.W.; Gartner, M.J.; Jarvik, R.K.; Weiss, W.J. The National, Heart, Lung, and Blood Instituted Pediatric Circulatolory Support Program: A Summary of the 5-year Experience. *Circulation* **2011**, *123*, 1233–1240. [CrossRef] [PubMed]
3. Miera, O.; Schmitt, K.R.; Delmo-Walter, E.; Ovroutski, S.; Hetzer, R.; Berger, F. Pump size of Berlin Heart EXCOR pediatric device influences clinical outcome in children. *J. Heart Lung Trans.* **2014**, *33*, 816–821. [CrossRef] [PubMed]
4. Lorts, A.; Zafar, F.; Adachi, I.; Morales, D.L. Mechanical Assist Devices in Neonates and Infants. *Semin. Thoracic Cardiovasc. Surg. Pediatric Cardiac Surg. Annu.* **2014**, *17*, 91–95. [CrossRef]
5. Gibber, M.; Wu, Z.J.; Chang, W.B.; Bianchi, G.; Hu, J.; Garcia, J.; Jarvik, R.; Griffith, B.P. In Vivo Experience of the Child-Size Pediatric Jarvik 2000 Heart: Update. *Asaio J.* **2010**, *56*, 369–376. [CrossRef]

6. Wei, X.; Li, T.; Li, S.; Son, H.S.; Sanchez, P.; Niu, S.; Watkins, A.C.; DeFilippi, C.; Jarvik, R.; Wu, Z.J.; et al. Pre-clinical evaluation of the infant Jarvik 2000 heart in a neonate piglet model. *J. Heart Lung Trans.* **2013**, *32*, 112–119. [CrossRef]
7. Baldwin, J.T.; Adachi, I.; Teal, J.; Almond, C.A.; Jaquiss, R.D.; Massicotte, M.P.; Dasse, K.; Siami, F.S.; Zak, V.; Kaltman, J.R.; et al. Closing in on the PumpKIN Trial of the Jarvik 2015 Ventricular Assist Device. *Semin. Thoracic Cardiovasc. Surg. Pediatric Cardiac Surg. Annu.* **2017**, *20*, 9–15. [CrossRef] [PubMed]
8. Allaire, P.E.; Kim, H.C.; Maslen, E.H.; Olsen, D.B.; Bearnson, G.B. Prototype Continuous Flow Ventricular Assist Device Supported on Magnetic Bearings. *Artif. Organs* **1996**, *20*, 582–590. [CrossRef] [PubMed]
9. Allaire, P.; Hilton, E.; Baloh, M.; Maslen, E.; Bearnson, G.; Noh, D.; Khanwilkar, P.; Olsen, D. Performance of a Continuous Flow Ventricular Assist Device: Magnetic Bearing Design, Construction, and Testing. *Artif. Organs* **1998**, *22*, 475–480. [CrossRef]
10. Locke, D.H.; Swanson, E.S.; Walton, J.F.; Willis, J.P.; Heshmat, H. Testing of a centrifugal blood pump with a high efficiency hybrid magnetic bearing. *Asaio J.* **2003**, *49*, 737–743. [CrossRef]
11. Merkel, T. Magnetic Bearing in INCOR Axial Blood Pump Acts as Multifunctional Sensor. In *Proceedings of Ninth International Symposium on Magnetic Bearings*; University of Kentucky Department of Mechanical Engineering: Lexington, KY, USA, 2004.
12. Wearden, P.D.; Morell, V.O.; Keller, B.B.; Webber, S.A.; Borovetz, H.S.; Badylak, S.F.; Boston, J.R.; Kormos, R.L.; Kameneva, M.V.; Simaan, M.; et al. The PediaFlow Pediatric Ventricular Assist Device. *Semin. Thoracic Cardiovasc. Surg. Pediatric Cardiac Surg. Annu.* **2006**, *9*, 92–98. [CrossRef] [PubMed]
13. Yumoto, A.; Shinshi, T.; Zhang, X.; Tachikawa, H.; Shimokohbe, A. *A One-DOF-controlled Magnetic Bearing for Compact Centrifugal Blood Pumps, Motion and Vibration Control*; Springer: Dordrecht, Netherlands, 2009.
14. Nguyen, Q.D.; Ueno, S. Analysis and Control of Nonsalient Permanent Magnet Axial Gap Self-Bearing Motor. *IEEE Trans. Ind. Electron.* **2011**, *58*, 2644–2652. [CrossRef]
15. Asama, J.; Hamasaki, Y.; Oiwa, T.; Chiba, A. Proposal and Analysis of a Novel Single-Drive Bearingless Motor. *IEEE Trans. Ind. Electron.* **2013**, *60*, 129–138. [CrossRef]
16. Takeda, R.; Ueno, S.; Jiang, C. Development of a Centrifugal Cryogenic Fluid Pump using an Axial Self-bearing Motor. In Proceedings of the 15th International Symposium on Magnetic Bearings, Kitakyushu, Japan, 3–6 August 2016.
17. Nojiri, C.; Kijima, T.; Maekawa, J.; Horiuchi, K.; Kido, T.; Sugiyama, T.; Mori, T.; Sugiura, N.; Asada, T.; Umemura, W.; et al. Development Status of Terumo Implantable Left Ventricular Assist System. *Artif. Organs* **2001**, *25*, 411–413. [CrossRef] [PubMed]
18. Maher, T.R.; Butler, K.C.; Poirier, V.L.; Gernes, D.B. HeartMate Left Ventricular Assist Devices: A Multigeneration of Implanted Blood Pumps. *Artif. Organs* **2001**, *25*, 422–426. [CrossRef]
19. Hoshi, H.; Shinshi, T.; Takatani, S. Third-generation Blood Pumps with Mechanical Noncontact Magnetic Bearings. *Artif. Organs* **2006**, *30*, 324–328. [CrossRef]
20. Farrar, D.J.; Bourque, K.; Dague, C.P.; Cotter, C.J.; Poirier, V.L. Design Features, Developmental Status, and Experimental Results with the Heartmate III Centrifugal Left Ventricular Assist System With a Magnetically Levitated Rotor. *Asaio J.* **2007**, *53*, 310–315. [CrossRef] [PubMed]
21. Timms, D.L.; Kurita, N.; Greatrex, N.; Masuzawa, T. BiVACOR A Magnetically Levitated Biventricualr Artificial Heart. In Proceedings of the 20th MAGDA conference in Pacific Asia, Kaohsiung, Taiwan, 14–16 November 2011.
22. Mehra, M.R.; Naka, Y.; Uriel, N.; Goldstein, D.J.; Cleveland Jr, J.C.; Colombo, P.C.; Walsh, M.N.; Milano, C.A.; Patel, C.B.; Jorde, U.P.; et al. A Fully Magnetically Levitated Circulatory Pump for Advanced Heart Failure. *N. Engl. J. Med.* **2017**, *376*, 440–450. [CrossRef] [PubMed]
23. Osa, M.; Masuzawa, T.; Tatsumi, E. Miniaturized axial gap maglev motor with vector control for pediatric artificial heart. *J. JSAEM* **2012**, *20*, 397–403.
24. Kurita, N.; Ishikawa, T.; Saito, N.; Masuzawa, T. Basic Design of the Maglev Pump for Total Artificial Heart by using Double Stator Type Axial Self-bearing Motor. In Proceedings of the 15th International Symposium on Magnetic Bearings, Kitakyushu, Japan, 3–6 August 2016.
25. Osa, M.; Masuzawa, T.; Tatsumi, E. 5-DOF Control Double Stator Motor for Paediatric Ventricular Assist Device. In *Proceedings of ISMB13*; University of Virginia: Charlottesville, VA, USA, 2012.

26. Osa, M.; Masuzawa, T.; Omori, N.; Tatsumi, E. Radial position active control of double stator axial gap self-bearing motor for pediatric VAD. *Mech. Eng. J.* **2015**, *2*, 15-00105. [CrossRef]
27. Osa, M.; Masuzawa, T.; Saito, T.; Tatsumi, E. Miniaturizing 5-DOF fully controlled axial gap maglev motor for pediatric ventricular assist devices. *Int. J. Appl. Electromagnet Mech.* **2016**, *52*, 191–198. [CrossRef]
28. Osa, M.; Masuzawa, T.; Orihara, R.; Tatsumi, E. Magnetic suspension performance enhancement of ultra-compact 5-DOF-controlled self-bearing motor for rotary pediatric ventricular assist device. In Proceedings of the 11th International Symposium on Linear Drives for Industry Applications (LDIA), Osaka, Japan, 6–8 September 2017.

© 2019 by the authors. Licensee MDPI, Basel, Switzerland. This article is an open access article distributed under the terms and conditions of the Creative Commons Attribution (CC BY) license (http://creativecommons.org/licenses/by/4.0/).

Article

Optimal Magnetic Spring for Compliant Actuation—Validated Torque Density Benchmark [†]

Branimir Mrak [1,2,*], Bert Lenaerts [2], Walter Driesen [2] and Wim Desmet [1,3]

[1] Department of Mechanical Engineering, Katholieke Universiteit Leuven, 3001 Leuven, Belgium; wim.desmet@kuleuven.be
[2] MotionS core lab, Flanders Make, 3001 Leuven, Belgium; bert.lenaerts@flandersmake.be (B.L.); walter.driesen@flandersmake.be (W.D.)
[3] DMMS core lab, Flanders Make, 3001 Leuven, Belgium
* Correspondence: branimir.mrak@flandersmake.be
† This paper is an expanded version on: Mrak, B.; Lenaerts, B.; Driesen, W.; Desmet, W. Optimal Design of Magnetic Springs; Enabling High Life Cycle Elastic Actuators. In proceedings of the 16th International Symposium on Magnetic Bearings (ISMB16), Beijing, China, 13–17 August 2018.

Received: 18 January 2019; Accepted: 18 February 2019; Published: 22 February 2019

Abstract: Magnetic springs are a fatigue-free alternative to mechanical springs that could enable compliant actuation concepts in highly dynamic industrial applications. The goals of this article are: (1) to develop and validate a methodology for the optimal design of a magnetic spring and (2) to benchmark the magnetic springs at the component level against conventional solutions, namely, mechanical springs and highly dynamic servo motors. We present an extensive exploration of the magnetic spring design space both with respect to topology and geometry sizing, using a 2D finite element magnetostatics software combined with a multi-objective genetic algorithm, as a part of a MagOpt design environment. The resulting Pareto-optima are used for benchmarking rotational magnetic springs back-to-back with classical industrial solutions. The design methodology has been extensively validated using a combination of one physical prototype and multiple virtual designs. The findings show that magnetic springs possess an energy density 50% higher than that of state-of-the-art reported mechanical springs for the gigacycle regime and accordingly a torque density significantly higher than that of state-of-the-practice permanently magnetic synchronous motors.

Keywords: magnetic spring; optimal design; component benchmarking; compliant actuation; parallel elastic actuators (PEA); series elastic actuators (SEA)

1. Introduction

The principles of elastic actuation, first introduced by Alexander et al. [1], whether in series [2] or in parallel [3] elastic actuators have been consistently proven to improve actuator performance in service robotics. These systems rely on the high torque and force density of mechanical springs to reduce peak power requirements and to improve the actuator's energy efficiency. For example, in work done by Mettin et al. [4], the energy consumption is reduced by 55%. The goal of this paper is to offer a robust spring solution, in the form of magnetic springs, that can extend the use of elastic actuation from service robotics to widespread industrial robots but also a much broader family of highly dynamic industrial motion systems.

A mechanical spring stores energy as the potential energy of elastic deformation. Spring design for highly dynamic loads in industrial use is typically limited by the long lifetime requirements and often leads to suboptimal designs for the purposes of elastic actuation. Conventionally, it was considered that for some metals there is a stress level called the fatigue limit, that can be sustained with an infinite lifetime [5]. Nowadays, this value is still often used in the design together with the stochastic design methods. However, the existence of a fatigue limit has been disputed even in the lab environment due to inclusions in the crystal lattice [6] of steel. Local stresses can lead to fatigue in any kind of metallic springs [5–8] and industrial environments impose additional risks (i.e., corrosive environment, temperature variations, mechanical handling, manufacturing limitations etc.). Often, high safety factors are employed to guarantee a robust design for a full product line, leading to heavy springs with high inertia.

Although the functionality of the magnetic spring (Figure 1) can be compared to that of a mechanical spring, the underlying physical principles are utterly different. Magnetic springs store potential energy in the magnetic field of permanent magnets (PM), where no fatigue failure mechanism is involved and thus have a virtually infinite lifetime [9], assuming the device is properly designed. This allows the use of compliant actuation concepts [10] in highly dynamic industrial applications with stringent lifetime demands.

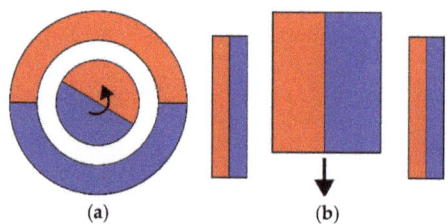

Figure 1. Conceptual drawing of (**a**) a rotational motion (torsional) magnetic spring and (**b**) linear motion (translational) and figure indicating torque and force generated due to the displacement.

With elastic actuators, it is possible to deliver more mechanically reactive power to the system, under the assumption of a higher torque density of springs compared to motors. Considering the evident benefit of using mechanical springs in service robotics in improving dynamic behavior, it is necessary to prove that magnetic springs have the same or higher energy density than conventional solutions with mechanical springs, in order to showcase their potential for the design of industrial motion systems.

Some of the target applications are torque oscillation, compensation in continuous rotation in internal combustion engines and windmills, reciprocating and intermittent motion in weaving looms [11], fast switching valves [12] (valvetrains in internal combustion engines), reciprocating pumps and compressors [13] and other tools and machines with a highly dynamic reciprocating motion. Additionally, magnetic springs have been reported for use in vibration reduction and vibration isolation [14] as well as for static load compensation [15]. It is worth mentioning that magnetic springs are topologically identical to passive magnetic bearings (PMB) and magnetic clutches. The main difference is the magnetic load point of the permanent magnets: in a magnetic spring the magnets are loaded over the entire B-H curve in each loading cycle, while for PMB and clutches the operating point remains constant for a constant mechanical load.

Unlike the previous efforts on the topic, where effort was focused on a specific use, this paper studies the optimal design of a magnetic spring in more detail and demonstrates systematically the impact of a magnetic spring on the performance of highly dynamic industrial actuators. This article is based on conference paper [16] where the optimal component design methodology was presented. That methodology was extended with a more elaborate, reproducible validation campaign including dynamic validation data, and multiple virtual optimal design points in requirement space. Additionally, for the purposes of benchmarking, more stress was put on the experimental validation and virtual validation using models of differing complexity. For the same reason, the mechanical and magnetic spring, including temperature effects on both the magnetic and mechanical springs, is considered. The closed form magnetic spring scaling model with model limitations is presented, as opposed to the intuitive yet incomplete model in the conference paper, making the experimental validation fully reproducible. Finally, in the discussion section, there is a significant amount of new benchmarking data for torque density comparison of magnetic springs and permanent magnet synchronous motors (PMSM). In addition, the exact data points and the designs will be made available via a link or in the addendum.

2. Materials and Methods

Within this article, the focus is primarily on the component design cycle but we will also present its complementarity with the system design cycle (Figure 2). The main subject of this study is rotational magnetic springs, although some of the considerations regarding energy density can also be translated to linear magnetic springs. Regarding the environment where the magnetic springs can be used, we consider that due to the limitations of permanent magnetic materials, environments where PMSM can operate are considered to be suitable for magnetic springs.

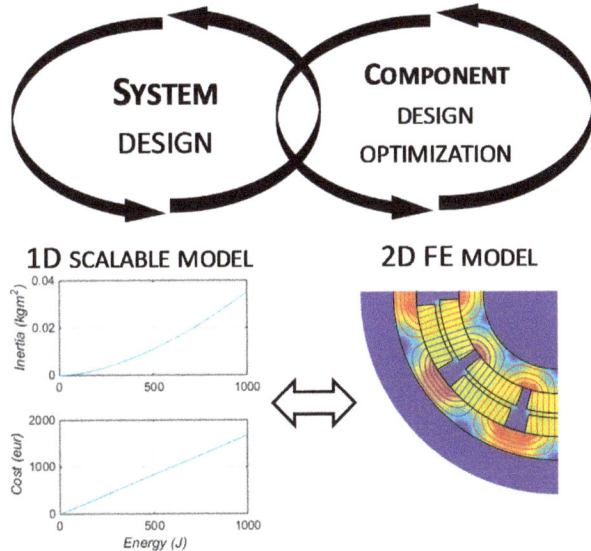

Figure 2. The co-dependent nature of the system design and component design cycles through linked modelling approaches.

Although the FE model of a detailed design geometry is an indispensable tool for component design, in system optimization the computational cost of finite elements (FE) can be prohibitively expensive. On the other hand, a scalable 1D dynamic model of a magnetic spring is the ideal model for the sizing of different drivetrain components and system optimization. Therefore, we define a 1D scalable model based on first principles, where cost and inertia of a magnetic spring are calculated directly from required reactive energy. This model can be iteratively updated based on the FE model results, as a result of the virtual validation where a 1D model is compared to optimal component designs coming from component optimization design.

A standard way to compare energy-storing devices is a Ragone chart [17]. It typically shows the tradeoff between energy density and power density, i.e., some energy storage components should be used when a high energy density is required (e.g., Li-ion batteries) and others when high instantaneous power is required (supercapacitors, flywheels). The bottleneck of such a static approach when it comes to highly dynamic drivetrains is the disregard for lifetime and system dynamics. In the highly dynamic applications targeted within this study, the mechanical power delivered to the system is significantly limited by the torque density of the actuator i.e., the ratio of torque limitation and inertia of the said actuator.

Therefore, it is necessary to know the inertia of the spring alongside the torque characteristics. Phenomenologically we can analyze the energy density of a spring. For elastic springs this will be the surface under the stress-strain curve (for the linear elastic model where the relation of stress and strain is linearly described by Young's modulus). Equivalently, for an idealized magnetic spring, the energy density is equivalent to the surface under the BH curve (Figure 3), which can be calculated as

$$E_{max} = \int_{H_c}^{0} B(H) dH \approx 2BH_{max} \qquad (1)$$

where B is the magnetic flux density and H is magnetic field strength.

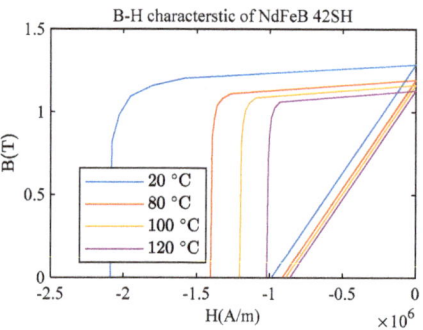

Figure 3. The potential energy of springs; permanent magnet energy density calculated from B-H characteristic is a measure of the maximum theoretical energy density of a magnetic spring.

Two assumptions about the magnetic spring have to be made in order to create a 1D scalable model. First of all, equal distribution of magnets between stator and rotor, resulting in perfect canceling of the magnetic field in the magnets in the case where maximum potential energy is stored within the magnet. Secondly, a fixed form factor of the rotor, following the 1st assumption and optimal rotor diameter achieved from FE simulation. It is important to note that variation of a permanent magnetic energy density of 30.79% for a 100 °C difference can result in a significant stiffness variation with temperature, while mechanical springs will normally have less than 5% [18] for the same region. Keep in mind that the magnetic springs are not expected to generate significant heat, yet, for an expected environmental variation of ±20 °C, the magnetic spring will have a variation of ±5%. Although the SmCo material has a higher Curie temperature and can allow for a slightly higher operational temperature, this large stiffness variation and the possibility of demagnetization limits the magnetic spring from operating in a high-temperature environment where mechanical springs face less severe limitations.

Furthermore, realistic designs of a magnetic spring will always have a lower energy density than the maximum theoretical limit, due to effects like fringing and flux leakage. Therefore, we can define the design efficiency as an energy density ratio of a realistic magnetic spring and an ideal magnetic spring

$$\eta_{mat} = \frac{E_{FE}}{E_{max}} \qquad (2)$$

and use it for 1D model correction based on FE results.

For the realistic embodiment of the magnetic spring concept, there is a range of feasible variants, both continuous (geometry sizing) and discrete (topological). By permutation of the discrete variants such as the PM materials in Table 1, or rotor and stator topologies shown in Figure 4, we can generate a number of topologies (Table 2), of which a number can be pruned out early in the design.

Table 1. Overview of considered permanent magnet materials.

Grade	N33H [1]	N42H [1]	S32H [2]	Pi-95HR [3]
Energy density (kJ/m^3)	521	673	510	173
BHmax (kJ/m^3)	263	334	255	85
Max temperature (°C)	120	120	350	125

[1] sintered Neodymium Iron Boron (NdFeB); [2] sintered Samarium Cobalt (SmCo), both anistropic material with limited magnetization; [3] plasto bonded NdFeB, isotropic material with free magnetization.

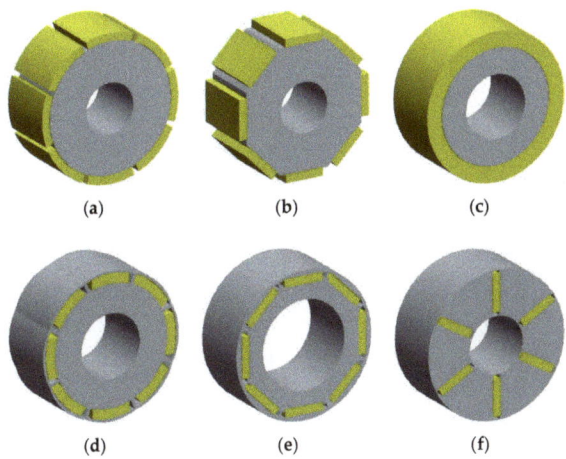

Figure 4. Overview of parametrized PM rotor topologies used in design optimization; (**a**) arc surface mounted magnets; (**b**) rectangular surface mounted magnets; (**c**) ring magnet—special case of arc surface mounted magnet; (**d**) buried arc magnets; (**e**) buried rectangular magnets; (**f**) internal magnets.

Table 2. Overview of evaluated topological choices.

Property	Variant
PM material	Isotropic, anisotropic temperature grade
Magnetization	Straight-diametrical, radial, tangential, Halbach
Magnetic array	Quasi-Halbach, multipole, over-segmented pole
Segment shape	Arc-segment, rectangular, bread loaf
Mounting method	Surface, buried, internal

For instance, surface mounted topologies are most suitable for achieving high torque density, and so are high energy density PM materials. However, in the case of PM materials, sintered NdFeB offer only limited magnetizations and are, as such, limiting in design options. The possibility to have more varied and better-suited magnetization is also why bonded rare-earth magnet solutions were studied. Of the listed topologies, the most promising were optimized and studied in more detail using MagOpt software [19].

When setting up the design specifications, it important to note that magnetic spring will not necessarily have a linear characteristic. In fact, except for small strokes around equilibrium positions, it is more likely to produce a quasi-sinusoidal characteristic. The above mentioned linear region can be extended by specific geometries of the magnet and back-iron. However, it has been noted that this can lead to lower design efficiency. Additionally, it is not a given fact that a linear characteristic is the most suitable solution for a given application case. An example of utilizing nonlinear spring can be found in Reference [20] where stable and unstable equilibria of magnetic spring can be used instead of a locking mechanism. Under this consideration, we need an alternative to spring stiffness to translate the system design specifications into component design specifications.

Specifying stroke and potential energy of a spring is adequate since it does not over-constrain the optimization problem by imposing a desired torque characteristic. The magnetic spring potential energy can be evaluated from torque characteristic and stroke as

$$E = \int_{\theta_1}^{\theta_2} T(\theta)d\theta. \qquad (3)$$

In order to evaluate each design variant, a 2D magnetostatics model of the geometry is calculated (Figure 5), for a range of θ sufficient to capture the desired rotational orders. Normally, odd higher orders, (3rd and 5th harmonic) are present for symmetric sine distortion. Therefore, in this analysis anywhere from 11 up to 21 θ points were used for a single design evaluation, with the lower numbers proven to be sufficient. For long rotors with the aspect ratio of length to diameter of more than two, the 2D approach should be sufficient, as cap effects can be disregarded. For disc geometries, on the other hand, it is necessary to use a 3D FEM. Since we are interested in high bandwidth actuators, it makes sense to focus on low inertia, long shaft solutions.

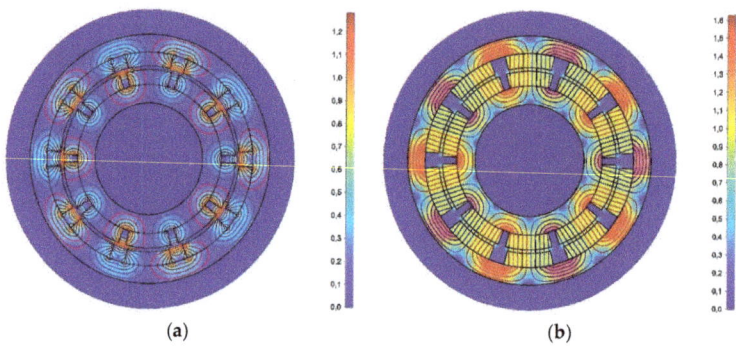

Figure 5. Overview of parametrized PM rotor topologies used in design optimization with surface mounted topologies a–c being most suitable for high torque density; (**a**) anti-aligned magnets resulting in unstable equilibrium with maximum energy stored in PM; (**b**) aligned magnets resulting in stable equilibrium; no energy stored in PM.

For each FEM evaluation, a list of metrics of interest can be calculated, either by pre-processing the specifications and the geometry or by post-processing the FEM solution. The considered design metrics are:

- Torque characteristic
    - Stored energy
    - Stroke
    - Higher harmonic content (Fourier decomposition/THD)
- Inertia
- Bulk material cost
- Demagnetization

The main objective of the design is to make a spring that fits the described energy and stroke specifications while minimizing inertia and cost. Within this article, the discussed cost of magnetic spring is merely the bulk material cost and is as such most useful for comparing different topological variations of magnetic springs, but also to get a first, rough idea of the magnetic spring cost in an industrial motion system. Although, in the latter case, other cost components, such as development, manufacturing and installation costs should be considered. All of these factors are heavily influenced by the volume of production and other economic factors. The cost comparison of different magnetic spring topological variations is considered valid, under the assumption that all of the considered sintered magnet geometries use the same manufacturing technology, especially with respect to magnetization i.e., only straight or diametrical magnetizations are considered. For a thorough design optimization, the MagOpt package [19] was used together with an opensource 2D FE solver for magnetostatic problems [21]. Other listed metrics were monitored for reasons of design safety (demagnetization) and possible unwanted dynamic effects (higher harmonic content). So far, loss models have not been considered, assuming that the efficiency of a magnetic spring is very high compared to a servo-drive since ohmic losses and the drive losses are completely avoided [22] in magnetic springs.

In order to validate the above described modeling approach, a prototype of a magnetic spring has been built (Figure 6). A magnetic spring consists of two diametrically magnetized ring NdFeB, N42H magnets, one on the stator and one on the rotor, with soft magnetic back iron to prevent flux leakage. For testing modularity, the spring has a built-in deep groove ball bearing. The bearing adds to the losses in the magnetic spring, which should be avoided in future designs, with a higher level of spring integration into the existing drivetrain.

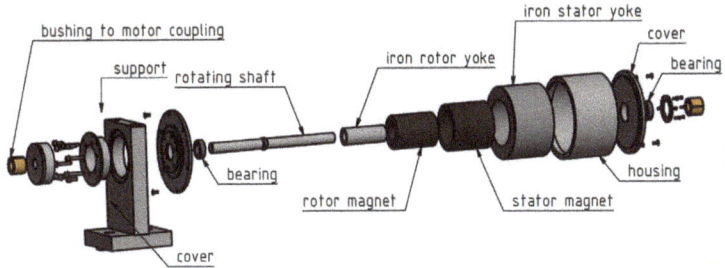

Figure 6. Explosion view of the prototyped magnetic spring design.

Experimental validation of the following modeling approach was conducted with a test rig, in Figure 7, that consists of a position controlled PMSM motor (1), driving an inertial wheel (3) with the assistance of spring (4). The torque sensors (2) are installed between the motor and the spring, and spring and the flywheel, using bellow couplings to avoid alignment issues or over-constrained rotation axes. Both dynamic and static experiments were conducted using the same setup. Note that here below couplings are adding serious elasticity in the system between the PMSM rotor and magnetic spring rotor but also the magnetic spring rotor and flywheel. This stiffness of the below couplings is, however, several orders of magnitude higher than that of the used magnetic spring and as such is not relevant for primary dynamics due to the reciprocating motion. For monitoring of power flows, the sensors (encoders and torque sensors), described in Table 3, are used together with fully observable controller inputs.

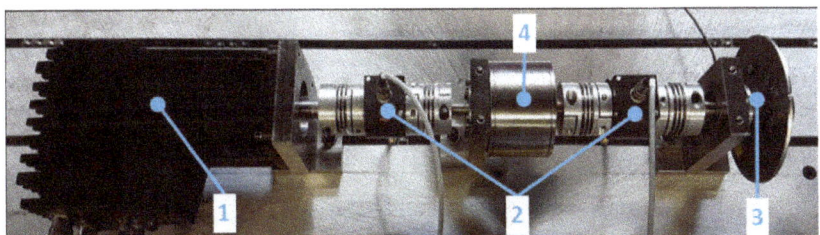

Figure 7. Experimental test rig consisting of a (1) position controlled PMSM, (2) torque sensors, (3) flywheel—load and the developed prototype of a magnetic spring (4).

Table 3. Experimental setup sensor specifications.

Location	Sensor Type	Range/Resolution
Motor −1	Integrated encoder motor	8192 pulse/rev
Flywheel −3	2× high accuracy optical encoder	327,680 pulse/rev
Spring −4		(14 bit with 40× interpolation)
Torque Sensors −2	Dynamic Torque Sensor	±100 Nm/±10 V

3. Results

3.1. Component Design Experimental Validation

The measurement results (Figure 8) show a good qualitative and quantitative fit of the static measurement and a good qualitative fit with respect to the low loss hypothesis. In Figure 8a, a slight skewing of the sinusoidal curve is visible. This phenomenon is related to the eccentricity of the magnetic center of design and the mechanical rotation axis due to the manufacturing tolerances and it can be captured in static stiffness modelled as a skewed sine due with single order eccentricity

$$T_{st}(\theta) = A \sin\left(\frac{2\pi\theta}{T} + \Delta T \sin\frac{\theta}{N}\right) \tag{4}$$

where T_{st} is the static torque of the magnetic spring as a function of angle θ. A is the torque amplitude in Nm. T is the period of the spring torque characteristic in radians, and depends on the pole pair number of the magnetic spring. ΔT, in radians, represents the skewing of the characteristic.

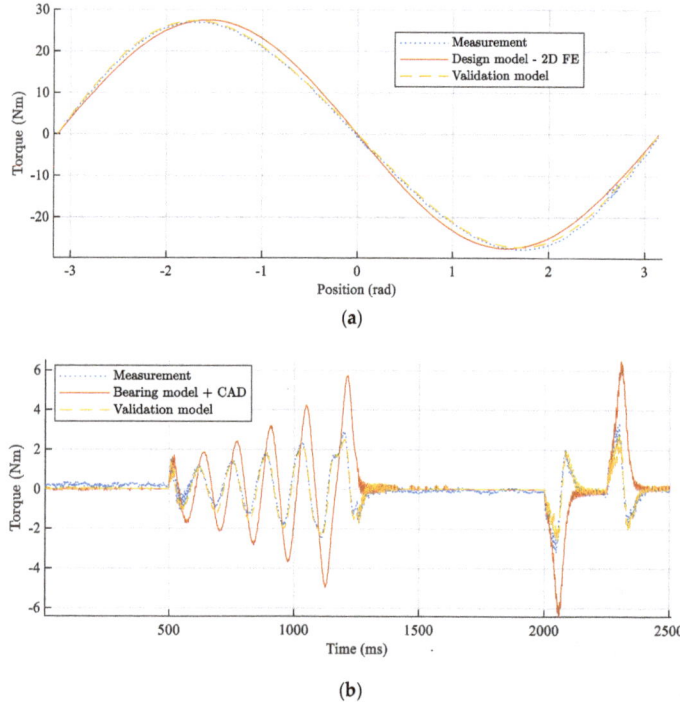

Figure 8. Component validation (a) static measurement of magnetic spring torque characteristic (b) dynamic measurement for identification of spring inertia and losses.

More complex aberration of the center of rotation can be captured with more addends in the sine argument, written as a Fourier decomposition, although, the most common issue of static alignment of mechanical and magnetic rotational axes results in a synchronous rotational order where $N = 1$. In Table 4, the identified parameters show that, apart from skewing, the peak torque value has less than 1% error compared to the FE model.

Table 4. Component design validation results.

Parameter	Measured/Estimated	Std. Error	Modeled Value
Inertia (kgm^2)	0.000420772	4.91751 × 10^{-6}	0.00042 [1]
Viscous friction coefficient (Nms/rad)	0.00195418	3.28588 × 10^{-4}	0.05 [2]
Coulomb friction torque (Nm)	0.126057	0.0156386	0.432 [3]
Torque Amplitude (Nm)	27.284	0.00757431	27.5
Skewing (rad)	0.132752	5.30995 × 10^{-4}	0

[1] CAD drawing of spring prototype; [2] based on bearing lubricant viscosity; [3] based on sliding torque; both from SKF model for W 61902-2Z bearing under C load.

For dynamic component identification (Figure 8b), torque and position measurements are filtered using 4th order 0-phase low pass filters with cutoff frequency at 200 Hz. The model parameters are fitted using Opti toolbox [Opti] non-linear least squares. The model of the simple magnetic spring in a direct dynamic form can be written as

$$T_{dyn} = T_{st}(\theta) - J\ddot{\theta} - c_v\dot{\theta} - T_C\left(\dot{\theta}\right), \tag{5}$$

where T_C, is dynamic Coulomb friction with hysteresis effect. The principal intrinsic losses of the magnetic spring are expected to be caused by the velocity proportional eddy currents in the permanent magnets. This way these can be set apart from the bearing friction that is dominated by a sliding and rolling friction that is constant above a certain speed (breakaway torque) and modeled as T_C. Additionally, viscous friction due to the lubricant viscosity, also contributes to bearing losses. The results of the dynamic parameter identification show that the losses are primarily dominated by the bearing friction T_C. Moreover, the speed proportional component is lower than the anticipated lubrication viscosity, proposed by the a priori bearing model [23]. Therefore, based on this experiment, it is impossible to discern between bearing losses and intrinsic magnetic spring losses. Nevertheless, a clear conclusion is that magnetic spring losses are negligible compared to the other energy sinks in the highly dynamic drivetrain.

The magnetic spring assisted drivetrain shown in Figure 7 can be operated between the unstable equilibria, similar to a parallel elastic actuator with a locking mechanism [3] or an inverted pendulum. The system is operating as follows (Figure 9). At $t = -0$ s, the load is held in a stable equilibrium. Initially, an FF torque pulse is applied together with a negative damping controller in order to excite the natural resonance of the system (phase 1. Start-up). Due to the negative damping, the load is slowly brought in the neighborhood of the unstable equilibrium where a stable PID controller is switched on in order to hold the load in position with 0-torque control (phase 2. 0-torque wait).

Once a reciprocating motion is required, the controller starts to operate in a catch-release fashion with a small FF torque pulse initiating the motion and pushing the load towards the next unstable equilibrium. Due to the magnetic spring torque, the load is accelerated until reaching the middle point, where the spring starts to decelerate it. Upon reaching the surroundings of the next unstable equilibrium, the motor is activated again, with a feedback controller, in order to stabilize the load in the endpoint. In this fashion, the motor is delivering only the bare minimum of the required torque.

The same motor operating without a magnetic spring while driving the same load (Figure 9), requires a peak torque of 25 Nm while in case of the magnetic spring assisted setup it is only 8 Nm. Therefore, the required peak torque is approximately three times lower in a case where a magnetic spring is used. The significant reduction can also be observed in energy consumption per cycle of reciprocating motion. The energy required for operation of magnetic spring assisted drivetrain is reduced from 29.07 J per cycle to 5.05 J per cycle, signifying an almost six-fold energy reduction. Here, the energy consumption is calculated as a sum of the measured mechanical power (torque sensors, encoders) at the motor output shaft and the ohmic losses calculated from the torque reference and motor datasheet parameters (phase resistance and torque constant).

It is visible that initially, during the start-up, the energy required to initialize the spring assisted setup is higher. This is, however, not a serious downside of the spring assisted actuator, considering that in the industrial application cases the drivetrain is only seldomly initiated, before long hours of operation, making the start-up energy consumption a negligible segment of the total energy consumption. For this reason, and for the convenience of tracking the energy consumption during the operational behavior (Figure 9, phase 3, reciprocating motion) the plotted energy is reset in the middle of the experiment (Figure 9, Phase 2, 0-torque wait. Alternatively, it is also possible to run the spring assisted system at a much higher torque in order to achieve a faster transient than it is possible with the motor only. In that case, a bang-bang controller can be used to accelerate the load as quickly as possible between two end positions.

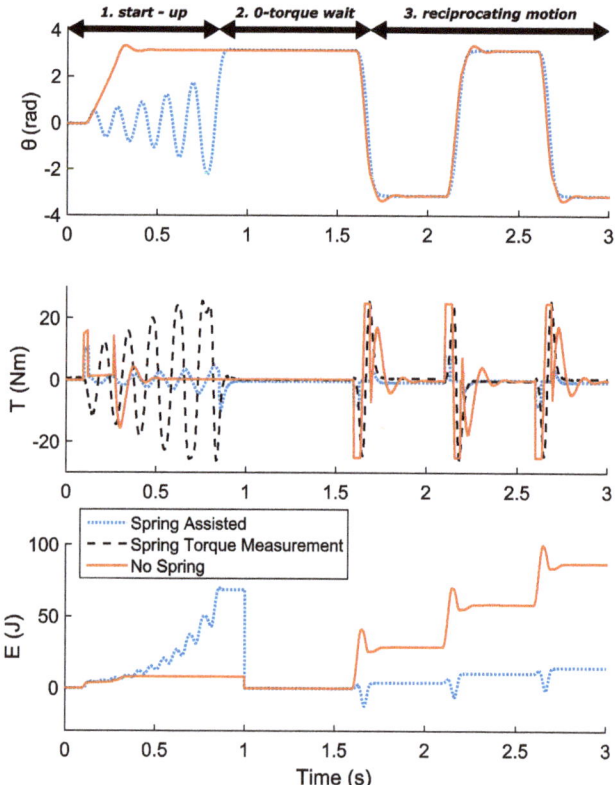

Figure 9. Proof-of-concept. Comparison of dynamic operational data for a magnetic spring with minimum motor torque vs. no spring setup with peak torque operation; controller structure and tuning have an impact on the exact values.

3.2. Model Based-Optimal Component Design

Detailed design optimization of the selected five most interesting topologies was done. The resulting Pareto fronts of different magnetic spring topologies can be compared for a fixed energy requirement and stroke. In Figure 10 it can be seen that sintered NdFeB is preferred over bonded magnets for reasons of both lower cost inertia. The added effect of using isotropic material (bonded NdFeB) to achieve a wider variety of magnetization is smaller than the added cost and inertia that results from lower flux densities in these materials. Interesting enough, low inertia levels can be achieved for each topology, irrelevant of the magnet geometry. However, the amount of material required to do so results in the lowest cost design with surface mounted arc magnets.

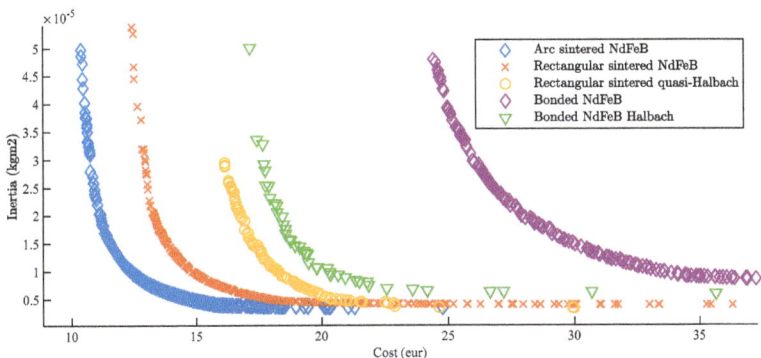

Figure 10. Optimization results plotted as Pareto-fronts for five stator and rotor topologies selected after design space pruning for different energy and stroke specifications.

Additional conclusions regarding design rules can be drawn from optimization results through Pareto optimal parameters. In Figure 11 normalized histograms (i.e., non- dimensional value on the y-axis) of the pareto optimal designs are plotted for each of the five selected topologies, showing the parameter distribution for the optimal designs that lie on the Pareto front.

Further analysis, shows that pole pitch in quasi Halbach arrays is optimally fully pitched with pitch factor values (i.e., the ratio of magnet coverage and pole pitch) approaching 1 (Figure 11e,f), which results in closest possible design to a real Halbach magnetization. On the other hand, standard multipole array values optimally have short pitchpoles with pitch factor values between 0.75 and 0.85 in order to prevent short-circuiting of the permanent magnet flux. The specific value of pitch factor, in this case, depends on the magnetic air gap between stator and rotor magnets as this represents the magnetic resistance of the parallel flux path. Another difference between Halbach and standard multipole arrays is in the thickness of the magnets (Figure 11c,d).

Finally, the scalable 1D model of magnetic springs can be validated using both the experimental validation and the detailed FEM of the designs presented here. Note that the two designs have different requirements as well as geometry sizing, and pole pair number. The single experimental design maps into one point, while the pareto front shows a dispersion of the possible designs. Therefore, the 1D model visible in Figure 12 Should be considered as a line partitioning the feasible component space from the infeasible.

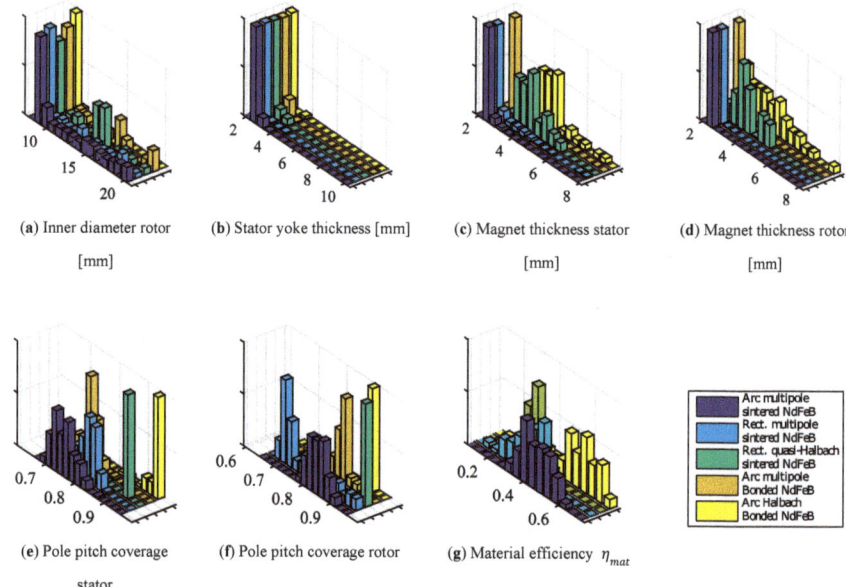

Figure 11. Optimal parameters histograms for five selected stator and rotor topologies selected after design space pruning.

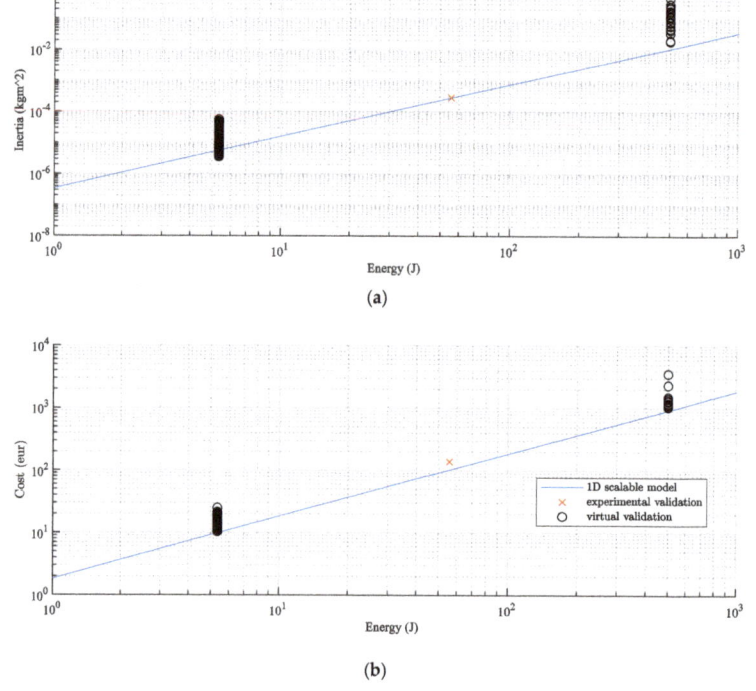

Figure 12. Validation of 1D scaling model for (**a**) magnetic spring inertia and (**b**) magnetic spring bulk material cost using experimental data and virtual validation data.

4. Discussion

Following the optimization results, the impact of magnetic spring on system performance can be analyzed from different perspectives. To compare magnetic springs to mechanical springs side-to-side phenomenologically, maximum theoretical energy density based on first principles is considered alongside with the realistically feasible energy density following from the optimization result. Since desired lifetime has a direct influence on stress level in mechanical springs and therefore also on energy density, we can plot energy density vs. required lifetime for mechanical and magnetic springs (Figure 13).

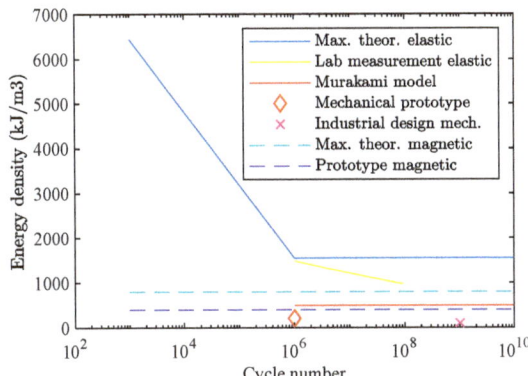

Figure 13. Benchmarking magnetic springs vs. mechanical springs; magnetic springs have increasingly higher energy density for high life cycle numbers.

The maximum theoretical energy density of the magnetic spring of $E_{max52S} = 828$ kJ/m³ is already higher than that of the Murakami model based gigacycle energy density of steel springs at $E_{Murakami} = 506$ kJ/m³. The difference between feasible energy density achieved with the feasible designs is even more dramatic. With NdFeB 42H grade and arc magnets, we are able to design a magnetic spring with an energy density of $E_{model42H} = 404$ kJ/m³, while a specific mechanical spring described in Reference [9] possesses an energy density of $E_{mech} = 210$ kJ/m³ with possible fatigue failure already at megacycles. However, it is difficult to generalize on feasible gigacycle mechanical designs for all the designs as the range of safety factors used within these applications is usually in quite a large range. Nevertheless, while being conservative we can say that the resulting increase in energy density is at least 50%, considering that the minimax design efficiencies of 0.6 achieved in this paper are larger than those in mechanical spring designs where safety factors are usually moderately higher than 2. Consequently, magnetic springs are specifically relevant for highly dynamic drivetrains in manufacturing machines e.g., a weaving loom operating shedding frames at 10 Hz reaches into megacycles after only 27.8 h and reaches well into a gigacycle regime in its standard operational age.

The benchmarking against PMSM is performed using a combination of real-life data from PMSM datasheets and model comparison using the developed magnetic spring modelling toolchain. Several types of servomotors are considered, with a preference for highly dynamic ones with high torque density. Extrapolation from the datasheet points can be carried out using the relation.

$$J = nT_{peak}^{5/3} \qquad (6)$$

which is valid for both springs and motors, assuming a fixed rotor aspect ratio (diameter/length). To reduce the cost and size of an electric drive solution, a reducer with transmission ratio n may be employed. However, the reflected inertia with a geared solution is always higher, given that

$$J = n^2 J_{motor} \tag{7}$$

and $2 > 5/3$. In Figure 14a only peak torques are considered, which are limited by the magnetic design of motor and spring. This results in a misleading image of rather "smaller" PMSMs (Maxon) having a higher torque density than magnetic springs with one or even two pole pairs. Note that for these "smaller" motors with natural cooling the difference between nominal torque and the peak torque is also greater. Figure 14b presents a more relevant image for highly dynamic industrial applications since here the nominal torque provided by the motor is compared to the spring peak torque.

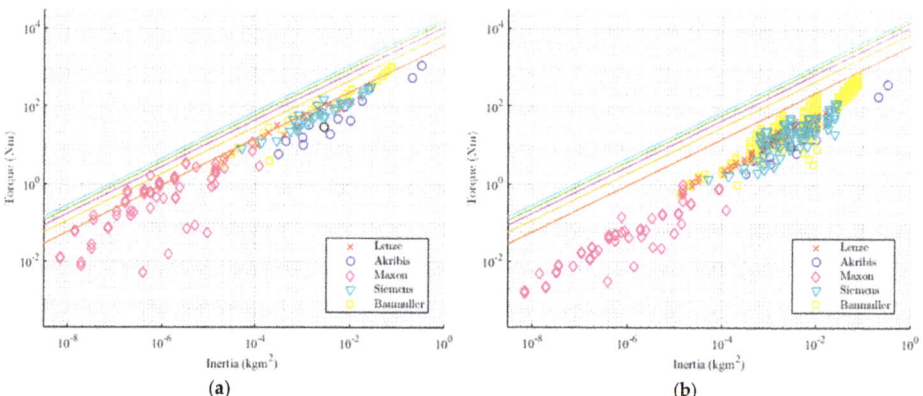

Figure 14. Benchmarking torque density of magnetic springs (1D scaling model) with pole pair numbers N_{pp} = 1–5 vs. off-the-shelf highly dynamic PMSM (**a**) peak torque vs. inertia and (**b**) nominal torque vs. inertia.

Nominal torque depends on the thermal design of the motor and cooling circuit, and in this analysis, all of the conventional air and liquid cooling methods were considered. The allowed dynamic peak can be higher than this limit, depending on the overload potential and the dynamic nature of the load, however, for exact quantification of this effect, a more detailed system dynamics analysis, outside of the scope of this article, should be considered. Additionally, for magnetic springs no thermal limitation is considered since the losses associated with generating torque are non-existent and the dynamic losses due to the eddy currents have been shown to be negligible. In conclusion, as a result of high torque density of magnetic springs the actuator bandwidth can be systematically improved for predetermined reciprocating profiles. This effect is more significant for small air-cooled motors, and less pronounced for larger designs. For exact quantification, a more detailed study on system optimization of magnetic spring assisted drivetrains is needed, considering optimal control strategy, relative sizing of spring and motor, sizing and selection of other system components (e.g., gearbox, motion conversion mechanism), and capturing motion requirements.

5. Conclusions

A detailed component design methodology has been developed and validated. A theoretical energy density was established based on physical insight in energy stored in permanent magnets. Detailed design optimization results show that design up to 60% of material efficiency are manufacturable. Best results are achieved with surface mounted arc sintered NdFeB magnets. The modeling approach is validated for a manufactured prototype, by static and dynamic component characterization on the experimental test rig. Following these results, based on 1D scalable models of magnetic springs, the energy density of a mechanical and magnetic spring can be compared for a long lifetime. Magnetic springs have at least a 50% higher power/energy density than mechanical springs with the with the added benefit of no fatigue failure. Additionally, using 1D scalable models of magnetic springs, a comparison between torque density of magnetic springs and PMSM off-the-shelf motors shows the added value of magnetic springs for preplanned reciprocating motion systems. The added benefit is specifically dramatic for partial strokes when magnetic springs with two and more pole pairs are employed when drivetrain peak acceleration is increased by 33% for the worst case scenario.

Conceptually, also the impact of magnetic spring on system behavior is experimentally demonstrated. The results show a six times lower energy consumption, and three times lower peak torque for a magnetic spring assisted drivetrain. Future studies should, however, consider a detailed analysis of system level design of highly dynamic drivetrains and quantify the associated cost reduction resulting from possible motor downsizing and improvement in bandwidth and energy efficiency in a more systematic manner.

Based on the dynamic measurement performed on the prototype, magnetic spring losses do not seem to be a relevant issue for the design of spring assisted reciprocating drivetrains, where due to the low pole pair number, the frequency of the magnetic field is rather low. Nevertheless, the demagnetization H_{demag} field is directly influenced by the temperature, and the rise in temperature is directly caused by losses. Therefore, it is important to notice that for using magnetic spring with higher pole pair number, than considered here, for e.g., torque ripple reduction [24], demagnetization can still be a possible issue. For such cases, a better understanding of thermal behavior and losses might lead to savings related to the selection of lower temperature grade magnets related to lower Dysprosium content.

Finally, we would like to comment un utility of magnetic springs. In this article, magnetic springs are primarily intended for enabling elastic actuation in industrial applications, where this was not feasible so far, due to the catastrophic failures that can result not only in significant down times, but also damage to the machine, processed goods and operator (e.g., weaving loom or punching tool failure). Alternatively, it might be possible for magnetic spring to replace mechanical springs in applications where the use of mechanical springs is established, for reasons of reduced downtime. However, the cost of magnetic springs is still expected to be higher than that of highly commoditized mechanical springs, meaning that a trade-off study, done from a perspective of specific industrial application will be necessary in order to determine when to use the magnetic springs. The results presented in this article provide a head start in such a study.

Author Contributions: Conceptualization, B.M., W.D. (Walter Driesen) and W.D. (Wim Desmet); Data curation, B.M.; Formal analysis, B.M.; Funding acquisition, W.D. (Walter Driesen) and W.D. (Wim Desmet); Investigation, B.M.; Methodology, B.M., B.L., W.D. (Walter Driesen) and W.D. (Wim Desmet); Supervision, W.D. (Walter Driesen) and W.D. (Wim Desmet); Validation, B.M.; Visualization., B.M.; Writing – original draft, B.M.; Writing – review & editing, B.M., B.L., W.D. (Walter Driesen) and W.D. (Wim Desmet).

Funding: The research of B. Mrak as an Early Stage Researcher was funded by a grant within the European Project EMVeM Marie Curie Initial Training Network (GA 315967). This research was partially supported by Flanders Make, the strategic research centre for the manufacturing industry within Flanders Make project Profensto_icon.

Acknowledgments: The authors would like to thank Linz Center of Mechatronics GmbH and Siegfried Silber for their support through the use of MagOpt software and the associated infrastructure.

Conflicts of Interest: The authors declare no conflict of interest.

References

1. Alexander, R.M. Three Uses for Springs in Legged Locomotion. *Int. J. Robot. Res.* **1990**, *9*, 53–61. [CrossRef]
2. Paluska, D.; Herr, H. The effect of series elasticity on actuator power and work output: Implications for robotic and prosthetic joint design. *Robot. Auton. Syst.* **2006**, *54*, 667–673. [CrossRef]
3. Vanderborght, B.; Verrelst, B.; Van Ham, R.; Van Damme, M.; Lefeber, D.; Meira, Y.; Duran, B.; Beyl, P. Exploiting natural dynamics to reduce energy consumption by controlling the compliance of soft actuators. *Int. J. Robot. Res.* **2006**, *25*, 343–358. [CrossRef]

4. Mettin, U.; La Hera, P.X.; Freidovich, L.B.; Shiriaev, A.S. Parallel Elastic Actuators as a Control Tool for Preplanned Trajectories of Underactuated Mechanical Systems. *Int. J. Robot. Res.* **2010**, *29*, 1186–1198. [CrossRef]
5. Bathias, C.; Paris, P.C. *Gigacycle Fatigue in Mechanical Practice*; CRC Press: Boca Raton, FL, USA, 2004; ISBN 9780203020609.
6. Puff, R.; Barbieri, R. Effect of non-metallic inclusions on the fatigue strength of helical spring wire. *Eng. Fail. Anal.* **2014**, *44*, 441–454. [CrossRef]
7. Abe, T.; Furuya, Y.; Matsuoka, S. Gigacycle fatigue properties of 1800 MPa class spring steels. *Fatigue Fract. Eng. Mater. Struct.* **2004**, *27*, 159–167. [CrossRef]
8. Serbino, E.M.; Tschiptschin, A.P. Fatigue behavior of bainitic and martensitic super clean Cr-Si high strength steels. *Int. J. Fatigue* **2014**, *61*, 87–92. [CrossRef]
9. Reinholz, B.A.; Seethaler, R.J.; Electromechanical, A. Experimental Validation of a Cogging-Torque-Assisted Valve Actuation System for Internal Combustion Engines. *IEEE/ASME Trans. Mechatron.* **2016**, *21*, 453–459. [CrossRef]
10. Sudano, A.; Tagliamonte, N.L.; Accoto, D.; Guglielmelli, E. A resonant parallel elastic actuator for biorobotic applications. In Proceedings of the IEEE International Conference on Intelligent Robots and Systems, Chicago, IL, USA, 14–18 September 2014; pp. 2815–2820.
11. Mrak, B.; Driesen, W.; Desmet, W. Magnetic Springs—Fast Energy Storage for Reciprocating Industrial Drivetrains. In Proceedings of the ABCM International Congress of Mechanical Engineering, Rio de Janeiro, Brazil, 17–21 May 2015; Volume 23.
12. Patt, P.J. Design and testing of a coaxial linear magnetic spring with integral linear motor. *IEEE Trans. Magn.* **1985**, *21*, 1759–1761. [CrossRef]
13. Poltschak, F.; Ebetshuber, P. Design of Integrated Magnetic Springs for Linear Oscillatory Actuators. *IEEE Trans. Ind. Appl.* **2018**, *54*, 2185–2192. [CrossRef]
14. Mizuno, T.; Takasaki, M.; Kishita, D.; Hirakawa, K. Vibration isolation system combining zero-power magnetic suspension with springs. *Control Eng. Pract.* **2007**, *15*, 187–196. [CrossRef]
15. Boisclair, J.; Richard, P.; Lalibert, T. Gravity Compensation of Robotic Manipulators Using Cylindrical Halbach Arrays. *IEEE/ASME Trans. Mechatron.* **2017**, *22*, 457–464. [CrossRef]
16. Mrak, B.; Lenaerts, B.; Driesen, W.; Desmet, W. Optimal Design of Magnetic Springs; Enabling High Life Cycle Elastic Actuators. In Proceedings of the 16th International Symposium on Magnetic Bearings ISMB16, Beijing, China, 13–17 August 2018.
17. Christen, T.; Ohler, C. Optimizing energy storage devices using Ragone plots. *J. Power Sources* **2002**, *110*, 107–116. [CrossRef]
18. Spittel, M.; Spittel, T. 4.2 Young's modulus of steel. In *Metal Forming Data of Ferrous Alloys—Deformation Behaviour*; Warlimont, H., Ed.; Springer: Berlin/Heidelberg, Germany, 2009; pp. 85–88. ISBN 978-3-540-44760-3.
19. Silber, S.; Koppelstätter, W.; Weidenholzer, G.; Bramerdorfer, G. MagOpt—Optimization tool for mechatronic components. In Proceedings of the 14th International Symposium on Magnetic Bearings, Linz, Austria, 11–14 August 2014; pp. 243–246.
20. King, C.; Beaman, J.J.; Sreenivasan, S.V.; Campbell, M. Multistable Equilibrium System Design Methodology and Demonstration. *J. Mech. Des.* **2004**, *126*, 1036. [CrossRef]
21. Meeker, D.C. Finite Element Method Magnetics, Version 4.2 (28Feb2018 Build). Available online: http://www.femm.info (accessed on 19 February 2019).
22. Wang, J.; Atallah, K.; Chin, R.; Arshad, W.M.; Lendenmann, H. Rotor eddy-current loss in permanent-magnet brushless AC machines. *IEEE Trans. Magn.* **2010**, *46*, 2701–2707. [CrossRef]
23. SKF Bearing Calculator. Available online: webtools3.skf.com/BearingCalc/ (accessed on 19 February 2019).
24. Mrak, B.; Adduci, R.; Weckx, S.; Driesen, W.; Desmet, W. Novel phase-bound magnetic vibration absorber for improved NVH performance of a wind turbine gearbox. In Proceedings of the International Conference on Noise and Vibration Engineering ISMA2018 Including USD2018, Leuven, Belgium, 17–19 September 2018.

 © 2019 by the authors. Licensee MDPI, Basel, Switzerland. This article is an open access article distributed under the terms and conditions of the Creative Commons Attribution (CC BY) license (http://creativecommons.org/licenses/by/4.0/).

Article

Condition Monitoring of Active Magnetic Bearings on the Internet of Things †

Alexander H. Pesch * and Peter N. Scavelli

Department of Engineering, Hofstra University, 104 Weed Hall, Hempstead, NY 11549, USA; PScavelli1@Pride.Hofstra.edu

* Correspondence: Alexander.H.Pesch@Hofstra.edu; Tel.: +1-516-463-7175

† This paper is an extended version of our paper published in: Pesch, A.H.; Scavelli, P.N. Condition Monitoring of AMBs on the IoT. In Proceedings of the 16th International Symposium on Magnetic Bearings (ISMB16), Beijing, China, 13–17 August 2018.

Received: 20 January 2019; Accepted: 14 February 2019; Published: 20 February 2019

Abstract: A magnetic bearing is an industrial device that supports a rotating shaft with a magnetic field. Magnetic bearings have advantages such as high efficiency, low maintenance, and no lubrication. Active magnetic bearings (AMBs) use electromagnets with actively controlled coil currents based on rotor position monitored by sensors integral to the AMB. AMBs are apt to the Internet of Things (IoT) due to their inherent sensors and actuators. The IoT is the interconnection of physical devices that enables them to send and receive data over the Internet. IoT technology has recently rapidly increased and is being applied to industrial devices. This study developed a method for the condition monitoring of AMB systems online using off-the-shelf IoT technology. Because off-the-shelf IoT solutions were utilized, the developed method is cost-effective and can be implemented on existing AMB systems. In this study, a MBC500 AMB test rig was outfitted with a Raspberry Pi single board computer. The Raspberry Pi monitors the AMB's position sensors and current sensors via an analog-to-digital converter. Several loading cases were imposed on the experimental test rig and diagnosed remotely using virtual network computing. It was found that remote AMB condition monitoring is feasible for less than USD 100.

Keywords: Active Magnetic Bearing; AMB; Internet of Things; IoT; Condition Monitoring

1. Introduction

Active Magnetic Bearings (AMBs) support a rotor with a magnetic field such that the rotor is levitated [1]. AMBs are an alternative to other types of bearings such as rolling element bearings and fluid film bearings [2]. AMBs have advantages over other types of bearings such as the potential for higher efficiency and rotational speeds. In addition, AMBs do not need lubrication, which is advantageous in clean rooms, and food or medical processing (e.g., [3]). AMBs do not need service and are useful in subsea [4] and space applications [5]. The control in the airgap clearance can be exploited, for example, for machining tool positioning [6] and active balancing [7].

Passive magnetic bearings use permanent magnet stators to repel a rotor with permanent magnets with opposing polarization [8]. The system is naturally stable, tending toward the equilibrium in the center of the bearing. Passive magnetic bearings tend to be more efficient than AMBs with electromagnets, because there is no current consumption. However, they lack the load capacity and performance of the actuated AMBs. Permanent-magnet-biased AMBs have permanent magnets to provide initial pulling force on a ferromagnetic rotor. Permanent-magnet-biased AMBs serve as a way to achieve some of the advantages of passive magnetic bearings and AMBs and still utilize integral sensors and actuators [9]. Zero-bias and low-bias AMBs also use integral sensors and high-efficiency actuators and require sophisticated control laws (e.g., [10–16]) to take advantage of the nonlinear flux.

AMBs use electromagnetic actuators to generate an attractive force on a ferromagnetic rotor. The magnitude of the attractive force increases as the rotor moves closer to the stator. Therefore, the setup is naturally unstable and requires stabilizing feedback control in order to function. An AMB inherently includes some form of position sensing. The rotor position signal is used to calculate required coil current to control rotor position and achieve stable levitation. A typical type of sensor is the noncontact eddy current position probe. AMB position sensors have conveniently been utilized for online health monitoring [17]. In addition, real-time knowledge of rotor position and coil current can be used to determine bearing forces that can be used to indirectly monitor rotor loading [18].

AMBs are designed, built, and implemented by engineers and technicians with expert knowledge. For example, AMB controller design must be customized to rotor geometry because of inertial and gyroscopic cross-coupling and to avoid excitation of flexible modes. Once commissioned on-site, the AMB can be used by the end user with relatively little training [19]. However, the end user may not be able to effectively troubleshoot complications with the AMB system that may arise after the commissioning process, because the end users are not trained to recognize or diagnose these complications. In such cases, a field service technician from the AMB's Original Equipment Manufacturer (OEM) must go on-site to perform service. This results in down time for the system supported by the AMB and increased expense to the customer. A solution is to make AMBs part of the Internet of Things (IoT). This would allow for increased productivity and decreased costs.

The exact definition and scope of the IoT is still being developed. However, the IoT basically enables the interconnection of physical devices. This interconnection allows the devices to send and receive data over the Internet. This enables value creation beyond the mere sum of the "thing-based function" and "IT-based service" [20]. By putting AMBs on the IoT, a remote user such as an off-site OEM technician could access the AMB, and diagnose malfunctions without the need for on-site examination. Therefore, (to reduce or eliminate equipment down time) the OEM technician can recommend corrective action immediately, or even preemptively.

Early cases of what would become known as IoT were in the area of radio-frequency identification (RFID) tags (e.g., [21]). Since then, there has been much development of IoT because of the significant impact on people's everyday lives [22]. There are several instances of industry beginning to take advantage of IoT technology [23]. More recently, IoT has been applied towards structural health monitoring [24]. For example, IoT has been used for monitoring the position of steel in a continuous casting process [25]. In addition, IoT has been used for the monitoring of vibrations in electric motors [26]. This suggests the potential of AMBs when coupled with the IoT.

There has been some previous work in the area of IoT tools used for AMBs. These are mostly proprietary industrial systems used toward facilitating automated commissioning. However, there has not been enough work in the application of off-the-shelf IoT hardware, which is low-cost and readily available. An early work in remote operation of AMBs is found in [27], where a local area network (LAN) is used to facilitate communication with real-time AMB controllers in a laboratory environment. The utilized LAN is a direct connection between the remote computer used for system interfacing and computers on the AMBs with dedicated hardware for real-time control and ethernet communications. This method is successful at interfacing with the AMBs for conducting experiments at a safe distance but was hardware intensive. In [28], a remote computer is used to communicate with a server computer via Transmission Control Protocol/Internet Protocol (TCP/IP). The server passes data via an RS-232 connection with a digital signal processor, which in turn controls an AMB system. The setup is used to remotely tune the AMB controller gains. Jayawant and Davies [29] developed an automated commissioning scheme capable of remote commissioning AMBs via TCP/IP and a Simple Object Access Protocol (SOAP) interface. SOAP sends packaged datasets between computers on a high level. Because the data are compiled and packaged before being transmitted via SOAP, the method can facilitate data transfer between different types of systems, e.g., differing operating systems. With SOAP, the actual data vector from the AMB is passed between the local and remote computers. Data processing can take place on either or both computers. In [30], SOAP-based remote commissioning is applied to a fluid film bearing AMB test rig and an industrial turbo-machine. Similar remote commissioning methods are utilized in [31] for a 3.3 MW motor-driven compressor and in [32] for a high-temperature gas-cooled reactor.

In the present study, an AMB test rig was augmented with an off-the-shelf IoT gateway that is low-cost and readily available. The local device was programmed to read the AMB's sensors and perform data processing for condition monitoring. A remote user could then log into the device through Virtual Network Computing (VNC) via TCP/IP to observe the sensor signals. This approach differs from the SOAP approach in that the AMB data vectors are not transmitted to the remote computer. In the current approach, all data processing is done on the local IoT gateway and the resulting frames are used remotely for condition monitoring. The usefulness of the developed system for condition monitoring of AMBs was demonstrated by operating the test rig under different conditions, and presenting the remote user interface, illustrating how the condition of the system was evaluated. Preliminary results for this study are presented in [33].

The next section explains the concept of using the IoT for condition monitoring of AMBs. Then, the experimental system used to demonstrate the proposed method is detailed, including cost information. Next, the experimental results are presented. The practical issue of sampling time when utilizing the IoT gateway device is then discussed. Finally, the paper is ended with concluding remarks.

2. Materials and Methods

This section covers the proposed method and the materials used to implement the experimental demonstration. Specifically, an introduction to AMBs and the proposed method for condition monitoring of AMBs on the IoT is discussed in the next subsection. Then, the AMB test rig for the experimental demonstration is presented. Finally, subsections for the hardware and software for IoT implementation are presented.

2.1. AMB and Condition Monitoring on the IoT

An AMB uses a magnetic field to support a rotating shaft. The magnetic field is generated by an array of electromagnets around the rotor. The electromagnetic force induced on the ferromagnetic rotor is inherently unstable. The position of the shaft is measured in real time by a noncontact position sensor. The position data are used (by a controller) to calculate how much current is needed in the electromagnetic coils to maintain a stable levitation. A typical control setup for AMBs is shown in Figure 1 [34]. Figure 1a shows a common biasing scheme for a single AMB axis. Figure 1b shows a control block diagram for a generic AMB-rotor system, which is Multiple-Input-Multiple-Output (MIMO) because a rotor may have multiple AMBs. Therefore, an AMB, by its nature, includes sensors and actuators. It is suitable for condition monitoring for traditional rotordynamic faults [35]. It is also highly apt to be extended to the IoT.

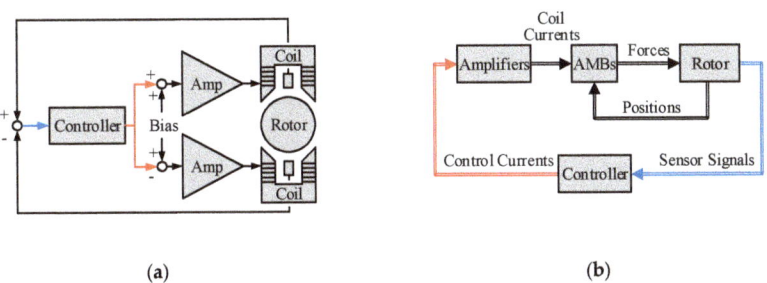

Figure 1. Typical AMB control scheme: (a) one control axis biasing; and (b) MIMO Control block diagram.

By accessing the information available in an AMB, a remote service technician can monitor the condition of the AMB system. Many types of information may be available in a AMB controller, such as rotor speed, hours in operation, temperature, flux, etc., but the signals used for the experimental demonstration in this work were position and current. Figure 2 illustrates the overall concept for the proposed method of AMB condition monitoring via the IoT. In the proposed scheme, the rotor position and the coil current, available from the AMB, are accessed for the IoT. Therefore, a service technician can log in remotely to the gateway and troubleshoot the AMB system. The remote technician is granted the ability to diagnose a variety of equipment malfunctions. The technician might be able to recommend corrective or even preventative action to the AMB end users without the need for an on-site service visit.

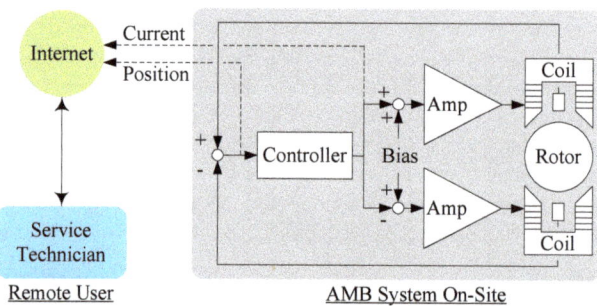

Figure 2. Condition monitoring of AMB system via the IoT concept.

For example, if the rotor position is low and/or the coil current is high, shaft overloading may be indicated. The technician can recommend checking the application, or "spec out" a larger AMB. Another scenario could involve the remote technician observing an overly large orbit indicating large unbalance. The technician could recommend rotor balancing before continuing operation.

2.2. Experimental System

The experimental system used to develop the proposed IoT condition monitoring solution consists of a stand-alone AMB test rig coupled with an IoT gateway and other hardware. The AMB test rig is model MBC500 by LaunchPoint Technologies, Inc. It has sensor signals that are readily available via a front breakout panel. The IoT gateway selected is the Raspberry Pi 3 Model B single board computer. The experimental system is shown in Figure 3. Figure 3 shows the overall system with: the AMB test rig, added sensor amplifiers and ADC unit on a solderless breadboard, and Raspberry Pi with Ethernet cable to connect to the Internet. The levitated shaft of the test rig has metal collars, which are movable, removable, made of differing materials for differing weights, and may have an adjustable unbalance screw added.

Additional hardware is required for interfacing the Raspberry Pi with the analog signals of the AMB rig. Specifically, an analog-to-digital converter (ADC) board and amplifiers were used to condition the sensor signals to the appropriate range. The following subsections detail the AMB test rig configuration, electronic hardware for IoT implementation, and corresponding software.

2.2.1. AMB Test Rig

The MBC500 AMB test rig consists of a single shaft supported by two radial AMBs at either end. Shaft position was monitored by Hall effect sensors to the immediate outside of the magnetic coils. The shaft is stainless steel and 12.5 mm in diameter. Rotation is driven by an air turbine near the inboard AMB.

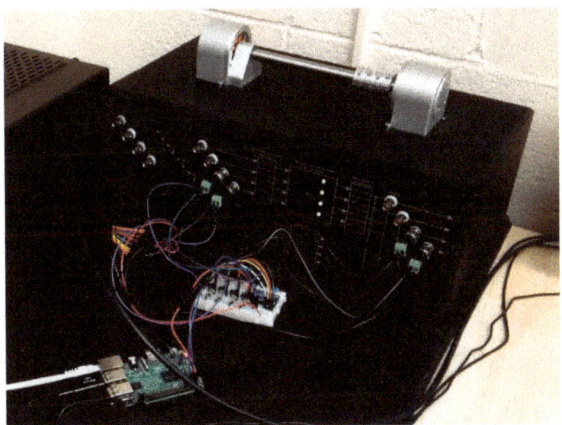

Figure 3. Experimental AMB test rig with attached Raspberry Pi single board computer, ADC, and conditioning circuitry.

Movable collars were added to the shaft to create reconfigurable weight and unbalance loads. The collars are approximately 10.5 mm wide and 28 mm in diameter. Two collars (one aluminum and one stainless steel) were used for the present test. The aluminum collar is 15 g and the stainless steel collar is 36 g. The position sensors are 2.8 mm from the ends of the shaft. The locations of the AMBs and the collars on the shaft are shown in Figure 4.

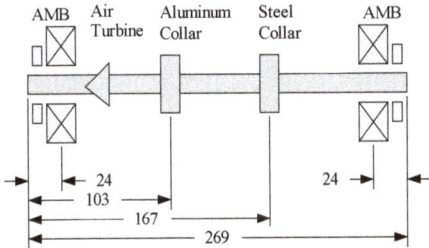

Figure 4. Rotor configuration. Dimensions in mm.

The AMBs have eight poles and are wired differentially in the vertical and horizontal axes. Each AMB axis has a bias current of 0.5 A and a nominal gap of 400 µm. The AMB force constant, based on coil geometry, is 2.8×10^{-7} N·m^2/A^2. The current amplifier bandwidth is approximately 720 Hz. The AMB controller built into the MBC500 is local lead-lag type. The built-in controller was used for the IoT condition monitoring study (when the shaft collars were moved).

2.2.2. Hardware Added for IoT

The IoT gateway selected is the Raspberry Pi 3 B (~USD 40). The Raspberry Pi is a single board computer with a 1.2 GHz Broadcom BCM2837 Quad-Core CPU and 1 GB of RAM. It runs the Raspbian operating system, which is Debian based. The Raspberry Pi was selected for this study as it is one of the most readily available IoT gateways. It serves as a cost-effective solution to remote condition monitoring. It is widely obtainable and therefore available to be augmented to older AMB systems already commissioned. In addition, software developed for the Raspbian operating system can be relatively easily ported to other Linux-based systems. A limitation of this IoT gateway solution is it is relatively slow and nonreal-time sampling as a consequence of the operating system. Therefore, it is most useful for monitoring relatively slow rotors or systems with low frequency bearing modes, subharmonics, external excitations, and substructure modes. The issue of nonreal-time sampling is discussed further in Section 3.

The hardware interface of the Raspberry Pi is general purpose input–output (GPIO) digital pins. To read the analog signals from the AMBs, an ADC must be added. For the current study, a Texas Instruments ADS1115 4-channel 16-bit ADC was utilized (~USD 15 on Adafruit Industries, LLC breakout board). Communication between the Raspberry Pi and the ADC was implemented via standard I^{2}C digital communication protocol. This protocol requires two wires, one for data transfer and one for a timing trigger. Figure 5a shows the basic scheme for experimental implementation of AMB condition monitoring via IoT.

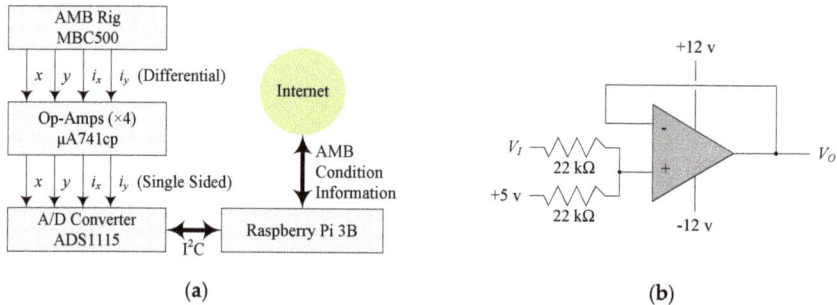

Figure 5. Hardware added for IoT implementation: (**a**) overall scheme; and (**b**) op-amp wiring diagram.

The ADC input range is nondifferential, effectively 0–5 V. The AMB sensors' operation range is within ±4 V. The sensor signals were scaled and offset by an array of four summing amplifiers. Texas Instruments µA741cp general purpose operational amplifiers (each < USD 1) were used. The amplifiers were wired as shown in Figure 4b to achieve the usable signal voltage range of 0.5–4.5 V. V_I is the input sensor signal from the AMB and V_O is the output signal sent to the ADC.

A standard PC power supply (~USD 20) provides an economical source of +12 V and −12 V, as well as the 5 V needed to raise to sensor signals. The resistors were balanced to half the overall sensor signal. The exact value of the resistors was selected through trial-and-error for an acceptable impedance match.

One AMB of the fully levitated system was monitored to demonstrate the developed IoT condition monitoring system. The sensor signals monitored were shaft position in the horizontal and vertical directions (x and y, respectively) and the corresponding axis coil currents (i_x and i_y, respectively). The software loaded on the Raspberry Pi used these data to give insight to the operating conditions of the outboard AMB. The entire assembly of hardware added for IoT is less than USD 100.

2.2.3. IoT Condition Monitoring Software

IoT condition monitoring software was written to collect the output signals of the ADC unit, and to a display a visual representation of the data. This includes rotor position and coil current. The ADC samples each signal, and converts the sampled data from voltage to bits. The bit readout was communicated to the Raspberry Pi via I^{2}C protocol. The bits were then scaled to recover the real-world signal values. This scaling involved applying an offset and a sensitivity to the vectors of collected data. The offset constants for the positions were obtained (during healthy levitation) by averaging the position vectors. These values were subtracted from all corresponding values in each corresponding vector. The sensitivity, found by comparing a known physical travel to the change in bits, was multiplied into each value in the respective vector. This yielded the calibrated position. A similar process was done for the current, but with the addition of offset to accurately represent the bias current applied.

Next, the vectors of position readings and current readings were plotted to display orbits (x vs. y) and position y with time and frequency. The frequency plot was generated by taking the Fast Fourier Transform (FFT) (via the NumPy library and the command *numpy.fft*) of a reconstructed position vector, as follows: the original sampled position vector had inconsistent time steps because of the nature of the Raspbian operating system. Since the processor was running both the operating system and the monitoring program, the loop running the code may be suspended to maintain the operating system's processes. (The exact sampling rate is discussed in Section 4.) To perform the FFT, which requires a consistent sampling rate, the original data vector and corresponding known time vector were resampled via linear interpolation. Using a resampled vector with 512 elements from the original approximately 400 over 5 s time history was selected. This number of virtual samples was selected to optimize the FFT algorithm while being similar to the actual number of data taken (to maintain fidelity).

The plotting of position and current in the time domain and in the frequency domain gave insight into the AMB system's operating condition. The software placed these figures, as well as the average value of each of the considered parameters after each instance, into a graphical user interface (GUI). The GUI was generated using the matplotlib library and the command *matplotlib.pyplot.plot*. This allowed taking the pre-allocated position and current vectors and expressing them visually. From the Raspberry Pi, a remote monitoring technician could observe the motions of the shaft within the bearing, and alert on-site operating technicians of a possible malfunction.

To connect remotely to the IoT AMB, Secure Shell (SSH), a cryptographic network protocol was enabled to allow a connection from an outside source. The Raspberry Pi hosted a server using the commercially available program *VNC Server*. The technician uses the corresponding program *VNC Viewer* (client) (both by RealVNC Ltd.) to facilitate the connection with a remote framebuffer (RFB) protocol. Similar IoT solutions utilizing VNC Server have been implemented in [36–38]. In general, an RFB protocol transmits screen pixels from one computer (over a network) to another and can also send control events, (e.g., mouse, keyboard, touch screen, etc.) in return [39]. For the current study, the RFB allowed the remote user to activate the IoT program as well as observe the operation of the bearing through the GUI. Therefore, not all data vectors for position, current, etc., need to be transmitted. The program performed data collection and displayed results for a set 5 s time interval. Future editions may allow the program to display latency, constantly update values and automatically replot figures.

3. Results

Six trials were conducted to demonstrate the condition monitoring capabilities of the developed AMB IoT system. For each trial, the shaft collars were adjusted to create varying load conditions. Two trials were conducted with the shaft levitated, but not rotating. Two trials were conducted with the shaft rotating. Two trials were conducted with the shaft rotating with added unbalance. For each case, the GUI used by the remote service technician is presented to illustrate how the condition of the AMB system is monitored.

3.1. Nonrotating Tests

Figures 6 and 7 show the GUI that a remote service technician would see when executing the IoT AMB monitoring software. The software was executed by calling VNC viewer on the local machine, connecting to the specific Raspberry Pi IoT gateway, and remotely calling the condition monitoring GUI program on the Raspberry Pi. The top of the GUI is a header that displays basic information. (The time period over which data are collected is displayed, in this case 5 s.) There is a blank expansion field for latency to be displayed by a later version of the software. The average values for position and current in the vertical and horizontal AMB axes are also displayed.

The two left plots are orbits, plotting data from vertical vs. horizontal AMB axes. The first plot displays rotor position and the second displays top coil current calculated from the recorded control current. The default scale for the position orbit is half the nominal AMB airgap of ±200 μm. The default scale serves as a limit for safe operation predetermined by expert users. Therefore, the technician can easily determine if the rotor is near the limit of safe operation if it nears one of the axes. The default scale for the current orbit is 0–1 A. A non-levitated rotor would sit at (0, −400) µm in the position orbit and (0,0) A in the current orbit.

The center right plot displays the time response of the rotor vertical position over the entire 5 s time history. The right plot displays the corresponding frequency spectrum found with an FFT of the resampled position data.

Figure 6 shows the results for the nonrotating shaft without added collars. The shaft levitated steadily near (0,0) µm, which is the center of the AMB. The coil current was near the bias current, 0.5 A. The calibration of the IoT condition monitoring system can differ from that of the AMB controller.

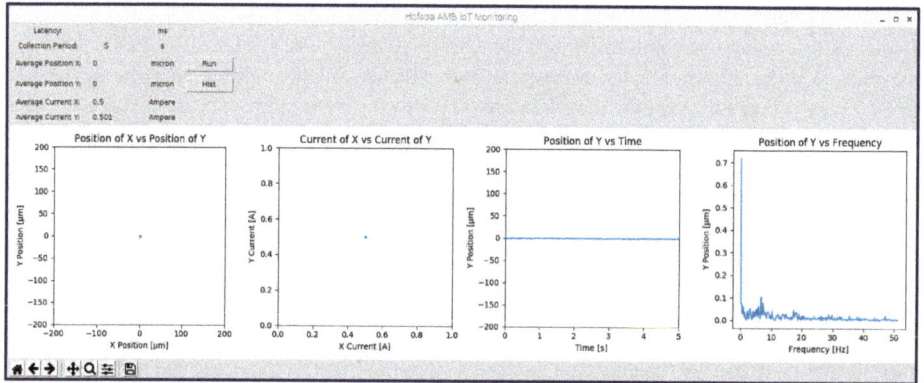

Figure 6. AMB IoT condition monitoring GUI display for non-rotating bare shaft.

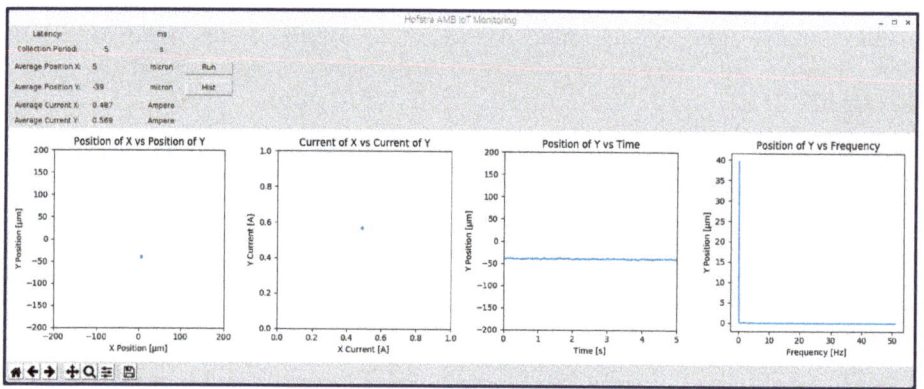

Figure 7. AMB IoT condition monitoring GUI display for non-rotating shaft with two collars.

The frequency spectrum depicted in the figure shows only a 0 Hz component for static offset. Note that the AMB controller has no integral action (as in a common PID controller). Figure 7 is for the nonrotating shaft with the two added collars. The remote service technician can infer the static loading condition of the levitated rotor by noting the lower levitated position and increased static current. In the event that the static deflection was too low or the static current was too high, the remote service technician can diagnose rotor over loading and recommend proper corrective action to on-site personnel without the need for an in-person inspection.

3.2. Balanced Rotating Tests

The rotor was rotated at 1200 RPM with and without the shaft collars. Figure 8 shows the IoT condition monitoring GUI for the case without the collars and Figure 9 for the case with the collars. The remote service technician can observe the orbit of the rotor inside the AMB air gap caused by the rotation. For both cases, the orbit was consistent and stable. The added weight of the shaft collars caused a static deflection downwards, and a corresponding increase of current (as with the nonrotating cases). The increase of gravity preloading also caused a slight bearing stiffness anisotropy, which led to vertical elongation of the orbit, which was observable by the remote service technician.

The rotation condition of the rotor was further observable in the time plot that displays a consistent harmonic wave. (The frequency spectrum had a peak at approximately 20 Hz, indicating the running speed.) The case with the shaft collars had a slightly higher peak at the running speed because of residual imbalance of the collars. The remote service technician can inspect the frequency spectrum for other components. For example, rotation off of the bearing centerline led to the appearance of a 2× rotation component at 40 Hz, as shown in Figure 9.

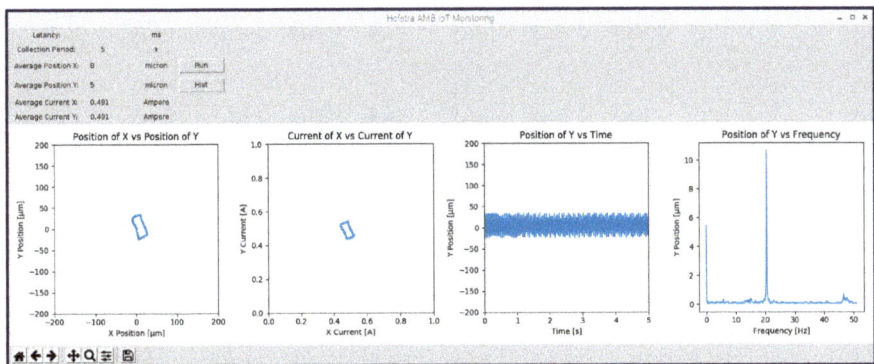

Figure 8. AMB IoT condition monitoring GUI display for rotating bare shaft.

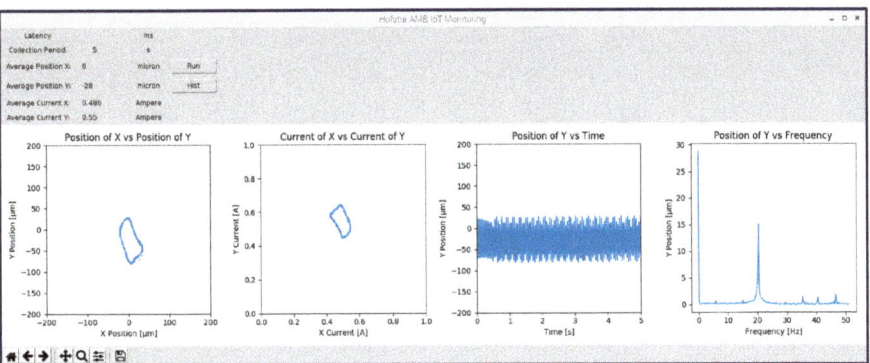

Figure 9. AMB IoT condition monitoring GUI display for rotating shaft with two balanced collars.

3.3. Unbalanced Rotating Tests

To induce a rotordynamic malfunction, unbalance masses were added to each shaft collar (in the form of a machine screw with exposed head). The resulting unbalance was approximately 6.5 g-mm per collar. Two unbalance tests were conducted. The first had both unbalance screws in the same direction on the rotor to create a static unbalance. The second had the unbalance screws in the opposite directions on the rotor to create a dynamic unbalance. Again, the shaft was rotated at 1200 RPM.

Figure 10 presents the IoT condition monitoring GUI for the static unbalance test and Figure 11 for the dynamic unbalance test. Observing the GUI in Figure 10, a remote service technician can diagnose the rotor unbalance from the slightly increased level of vibrations. This was seen in orbit size, vibration amplitude in time, and 1× frequency peak. The slight increase in level of vibration can alert the remote service technician to the added unbalance.

The dynamic unbalance test shows that the unbalance increased in the aluminum (inboard) collar but the added mass of the stainless steel (outboard) collar countered its own residual unbalance. Therefore, the remote service technician would observe a healthier orbit size, albeit lower, in the bearing gap. The ability to remotely access these data enables the remote service technician to recommend rotor balancing to the end user.

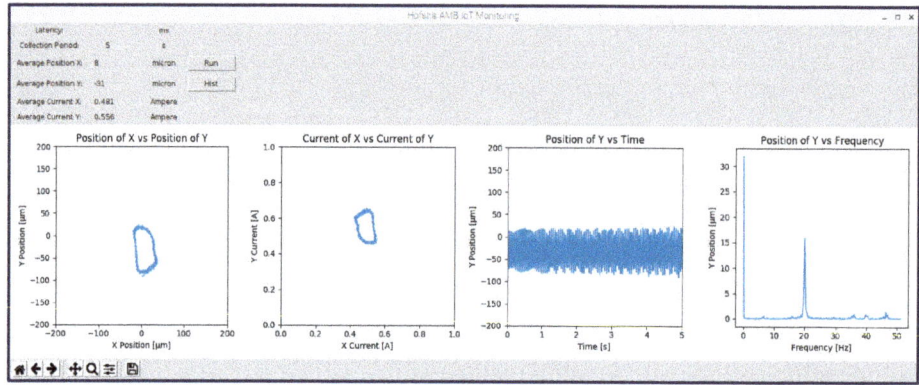

Figure 10. AMB IoT condition monitoring GUI display for rotating shaft with two collars and static unbalance.

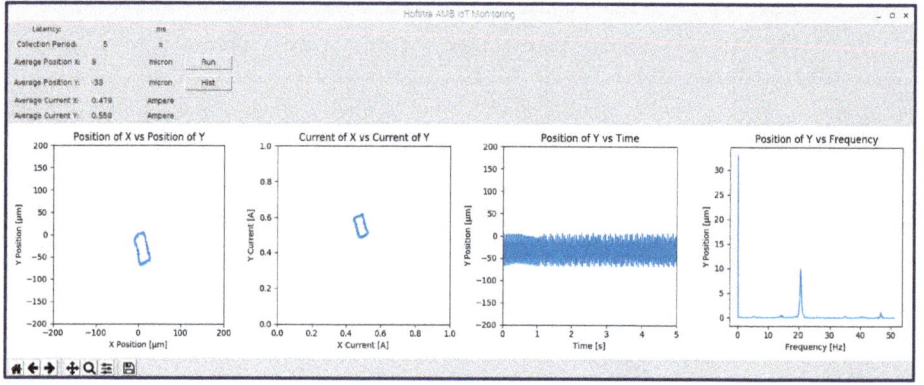

Figure 11. AMB IoT condition monitoring GUI display for rotating shaft with two collars and dynamic unbalance.

4. Discussion

A limitation of the developed IoT condition monitoring solution is the inconsistent sampling rate that stems from the operating system of the IoT gateway device. This is different from, for example, a dedicated microcontroller that has no operating system and no related background activities. The developed IoT program executed in Raspbian achieved a typical sampling rate of 100 Hz. However, it suffered from periodic delays created as the operating system performs background processes. These are the functions maintaining the operating system and functionality of peripherals and other programs

run by the remote user. Figure 12a shows the time stepping history of a characteristic 5 s condition monitoring run.

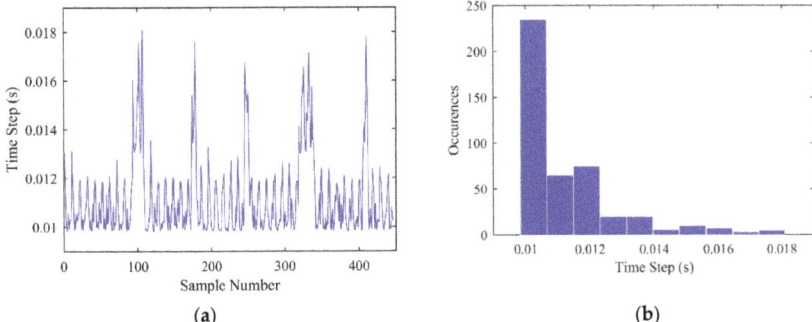

Figure 12. Characteristic sampling times for 5 s of IoT data: (**a**) time history; and (**b**) histogram.

The nominal sampling time of 0.01 s is presented as the baseline level in the figure. Frequently, the time was delayed to around 0.012 s. In addition, the data collection was pseudo-periodically delayed even further for several time steps, about every 1 s. This led to a time step as high as 0.018 s.

Figure 12b shows a histogram of the sampling time for this 5 s run. The histogram confirmed that the nominal sampling time was dominant, but interrupted by occasional delays. The current study overcame this limitation by visual inspection of orbits (which are not highly time dependent), and resampling to achieve a practical frequency spectrum. (However, this issue is important in further development of IoT for AMBs, i.e., execution of active control online.) A possible solution might be implementation of a real-time operating system. Another solution might be implementation of a programmable real-time unit on a single board computer.

5. Conclusions

This study addressed the problem of remote condition monitoring of AMBs. A solution was proposed to use off-the-shelf IoT hardware and custom software to tie into an AMB's position and current signals. This allowed an OEM technician to observe the signals remotely. The proposed strategy was demonstrated on an AMB test rig. A Raspberry Pi gateway and VNC Server software were used to implement IoT connection. The IoT gateway and other associated hardware cost less than USD 100. Static loading and static and dynamic unbalances were imposed on the experimental rotor. For each case, the conditions of the AMB system were successfully monitored remotely.

Therefore, it was concluded that off-the-shelf IoT hardware and custom software is economical and effective for remote AMB condition monitoring. AMB OEMs can implement similar methods to remotely monitor their products, which are operating on-site for their clients, the end users. This ability will alleviate the need for on-site service calls, and prevent AMB down time.

There are several promising directions for further development of AMBs and IoT. First, cybersecurity should be considered. In other words, mechanisms need to be developed to ensure only intended users can log in and access AMB data. In addition, a mobile application can be developed with which AMB users can check on the condition of the system from arbitrary locations. More complicated is the potential improvement of the IoT scheme for real-time use. Two possible solutions are a real-time operating system for the IoT gateway and using a real-time programmable unit, which would increase hardware cost. Real-time execution will lead to the next stage of development, a cyber physical system (i.e., the feedback control for the AMB will be done through the IoT, making a system in which the real-world dynamics of the system are dependent on the cyberworld). Then, an OEM service technician would not only be able to monitor the condition of an AMB system and diagnose problems, but also might be able to fix problems by changing the control law.

Author Contributions: The overall AMB condition monitoring scheme was devised by A.H.P. the coding for IoT implementation of the digital hardware was conducted by P.N.S. The authors worked together to perform the experiment.

Conflicts of Interest: The authors declare no conflicts of interest.

References

1. Bleuler, H.; Cole, M.; Keogh, P.; Larsonneur, R.; Maslen, E.; Okada, Y.; Schweitzer, G.; Traxler, A. *Magnetic Bearings: Theory, Design, and Application to Rotating Machinery*; Schweitzer, G., Maslen, E.H., Eds.; Springer: Berlin, Germany, 2009.
2. Pesch, A.H.; Sawicki, J.T. Active Magnetic Bearing Online Levitation Recovery through µ-Synthesis Robust Control. *Actuators* **2017**, *6*, 2. [CrossRef]
3. Osa, M.; Masuzawa, T.; Orihara, R.; Tatsumi, E. Compact Maglev Motor with Full DOF Active Control for Miniaturized Rotary Blood Pumps. In Proceedings of the 11th International Symposium on Linear Drives for Industry Applications, Osaka, Japan, 6–8 September 2017.
4. Aeschlimann, M.; Hubatka, M.; Stettler, R.; Housseini, R. Commissioning of Off-Shore Gas Compressor with 9-Axes Magnetic Bearing System: Commissioning Tools. In Proceedings of the ASME Turbo Expo 2017: Turbomachinery Technical Conference and Exposition, Charlotte, NC, USA, 26–30 June 2017.
5. Han, B.; Zheng, S.; Li, H.; Liu, Q. Weight-Reduction Design Based on Integrated Radial-Axial Magnetic Bearing of a Large-Scale MSCMG for Space Station Application. *IEEE Trans. Ind. Electron.* **2017**, *64*, 2205–2214. [CrossRef]
6. Smirnov, A.; Pesch, A.H.; Pyrhönen, O.; Sawicki, J.T. High-precision Cutting Tool Tracking with a Magnetic Bearing Spindle. *J. Dyn. Syst. Meas. Contr.* **2015**, *137*, 051017. [CrossRef]
7. Zheng, S.; Chen, Q.; Ren, H. Active Balancing Control of AMB-Rotor System using a Phase-Shift Notch Filter Connected in Parallel Mode. *IEEE Trans. Ind. Electron.* **2016**, *63*, 3777–3785. [CrossRef]
8. Lijesh, K.P.; Muzakkir, S.M.; Hirani, H. Failure Mode and Effect Analysis of Passive Magnetic Bearing. *Eng. Fail. Anal.* **2016**, *62*, 1–20. [CrossRef]
9. Filatov, A.; Hawkins, L.; McMullen, P. Homopolar Permanent-Magnet-Biased Actuators and Their Application in Rotational Active Magnetic Bearing Systems. *Actuators* **2016**, *6*, 26. [CrossRef]
10. Tsiotras, P.; Wilson, B.C. Zero- and Low-Bias Control Designs for Active Magnetic Bearings. *IEEE Trans Contr. Syst. Technol.* **2003**, *11*, 889–904. [CrossRef]
11. Tsiotras, P.; Arcak, M. Low-Bias Control of AMB Subject to Voltage Saturation: State-Feedback and Observer Designs. In Proceedings of the 41st IEEE Transactions on Control Systems Technology, Las Vegas, NV, USA, 11–13 December 2002.
12. Mystkowski, A.; Pawluszewicz, E.; Dragašius, E. Robust Nonlinear Position-flux Zero-bias Control for Uncertain AMB System. *Int. J. Contr.* **2015**, *88*, 1619–1629. [CrossRef]
13. Mystkowski, A.; Pawluszewicz, E. Nonlinear Position-flux Zero-bias Control for AMB System with Disturbance. *Appl. Comput. Electromagnet. Soc. J.* **2017**, *32*, 650–656.
14. Tsiotras, P.; Wilson, B.; Bartlett, R. Control of a Zero-bias Magnetic Bearing using Control Lyapunov Functions. In Proceedings of the 39th IEEE Conference on Decision and Control, Sydney, Australia, 12–15 December 2000.
15. Jastrzebski, R.P.; Smirnov, A.; Mystkowski, A.; Pyrhönen, O. Cascaded Position-Flux Controller for AMB System Operating at Zero Bias. *Energies* **2014**, *7*, 3561–3575. [CrossRef]
16. Mystkowski, A. Lyapunov Sliding-Mode Observers with Application for Active Magnetic Bearing Operated with Zero-bias Flux. *J. Dyn. Syst. Meas. Contr.* **2018**, *141*. [CrossRef]
17. Pesch, A.H. Damage Detection of Rotors using Magnetic Force Actuator: Analysis and Experimental Verification. Master's Thesis, Cleveland State University, Cleveland, OH, USA, 2008.
18. Auchet, S.; Chevrier, P.; Lacour, M.; Lipinski, P. A New Method of Cutting Force Measurement based on Command Voltages of Active Electro-magnetic Bearings. *Int. J. Mach. Tools Manuf.* **2004**, *44*, 1441–1449. [CrossRef]
19. Smirnov, A. AMB System for High-Speed Motors Using Automatic Commissioning. Ph.D. Thesis, Lappeenranta University of Technology, Lappeenranta, Finland, 2012.
20. Wortmann, F.; Flüchter, K. Internet of Things–Technology and Value Added. *Bus. Inf. Syst. Eng.* **2015**, *57*, 221–224. [CrossRef]

21. Ashton, K. Internet of Things. *RFID J.* **2009**, *22*, 97–114.
22. Atzori, L.; Iera, A.; Morabito, G. The Internet of Things: A Survey. *Comput. Netw.* **2010**, *54*, 2787–2805. [CrossRef]
23. Xu, L.D.; He, W.; Li, S. Internet of Things in Industries: A Survey. *IEEE Trans. Ind. Inf.* **2014**, *10*, 2233–2243. [CrossRef]
24. Tokognon, C.A.; Gao, B.; Tian, G.Y.; Yan, Y. Structural Health Monitoring Framework Based on Internet of Things: A Survey. *IEEE Internet Things J.* **2017**, *4*, 619–635. [CrossRef]
25. Zhang, F.; Liu, M.; Zhou, Z.; Shen, W. An IoT-Based Online Monitoring System for Continuous Steel Casting. *IEEE Internet Things J.* **2016**, *3*, 1355–1363. [CrossRef]
26. Ganga, D.; Ramachandran, V. IoT Based Vibration Analytics of Electrical Machines. *IEEE Internet Things J.* **2018**, *5*, 4538–4549. [CrossRef]
27. Hilton, E.F.; Humphrey, M.A.; Stankovic, J.A. Design of an Open Source, Hard Real Time, Controls Implementation Platform for Active Magnetic Bearings. In Proceedings of the 7th International Symposium on Magnetic Bearings, Zurich, Switzerland, 23–25 August 2000.
28. Turker, E.; Harvey, A. Magnetic Bearing Controller Tuning and Client/Server Technology. In Proceedings of the TENCON 2005–2005 IEEE Region 10 Conference, Melbourne, Australia, 21–24 November 2005.
29. Jayawant, R.; Davies, N. Integration of Signal Processing Capability in an AMB Controller to Support Remote and Automated Commissioning. In Proceedings of the 1st Brazilian Workshop on Magnetic Bearings, Rio de Janeiro, Brazil, 25–26 October 2013.
30. Jayawant, R.; Davies, N. Field Experience with Automated Tools in Both Remote and Local Commissioning of Active Magnetic Bearing Systems. In Proceedings of the 14th International Symposium on Magnetic Bearings, Linz, Austria, 11–14 August 2014.
31. Jayawant, R.; Masala, A. Design and Commissioning of a 3.3 MW Motor-driven Compressor Fully Supported on Active Magnetic Bearings. In Proceedings of the 15th International Symposium on Magnetic Bearings, Kitakyushu, Japan, 3–6 August 2016.
32. Guo, Z.; Zhou, G.; Shultz, P.R.; Qian, M.; Zhu, L. Design and Qualification Testing of Active Magnetic Bearings for High-temperature Gas-cooled Reactors. In Proceedings of the 15th International Symposium on Magnetic Bearings, Kitakyushu, Japan, 3–6 August 2016.
33. Pesch, A.H.; Scavelli, P.N. AMB Condition Monitoring on the IoT. In Proceedings of the 16th International Symposium on Magnetic Bearings, Beijing, China, 13–17 August 2018.
34. Anantachaisilp, P.; Lin, Z. Fractional Order PID Control of Rotor Suspension by Active Magnetic Bearings. *Actuators* **2017**, *6*, 4. [CrossRef]
35. Sawicki, J.T.; Friswell, M.I.; Pesch, A.H.; Wroblewski, A.C. Condition Monitoring of Rotor using Active Magnetic Actuator. In Proceedings of the ASME Turbo Expo. 2008: Power for Land, Sea, and Air, Berlin, Germany, 9–13 June 2008.
36. Mada Sanjaya, W.S.; Maryanti, S.; Wardoyo, C.; Anggraeni, D.; Aziz, M.A.; Marlina, L.; Roziqin, A.; Kusumorini, A. The Development of Quail Eggs Smart Incubator for Hatching System based on Microcontroller and Internet of Things (IoT). In Proceedings of the International Conference on Information and Communications Technology, Yogyakarta, Indonesia, 6–7 March 2018.
37. Philip, M.; Mendonca, P.J.; Thampy, A.; Menezes, M.; Tantry, R. IoT based Energy Meter (AMMP). *Int. J. Internet Things* **2017**, *6*, 88–90.
38. Nagamma, N.N.; Narmada, T.; Lakshmaiah, M.V.; Ramesh, V.; Pakardin, G. VNC Server based EVM System. In Proceedings of the International Conference on Energy, Communication, Data Analytics and Soft Computing, Chennai, India, 1–2 August 2017.
39. How does VNC Technology Work? Available online: https://www.realvnc.com/en/connect/docs/faq/function.html#how-does-vnc-technology-work (accessed on 4 January 2019).

© 2019 by the authors. Licensee MDPI, Basel, Switzerland. This article is an open access article distributed under the terms and conditions of the Creative Commons Attribution (CC BY) license (http://creativecommons.org/licenses/by/4.0/).

Article

Stability and Performance Analysis of Electrodynamic Thrust Bearings[†]

Joachim Van Verdeghem *, Virginie Kluyskens and Bruno Dehez

Department of Mechatronic, Electrical energy and Dynamic systems (MEED), Institute of Mechanics, Materials and Civil Engineering (IMMC), Université catholique de Louvain (UCLouvain), 1348 Louvain-la-Neuve, Belgium; virginie.kluyskens@uclouvain.be (V.K.); bruno.dehez@uclouvain.be (B.D.)
* Correspondence: joachim.vanverdeghem@uclouvain.be
† This paper is an extended version of our paper published in Van Verdeghem, J.; Kluyskens, V.; Dehez, B. Comparison Criteria and Stability Analysis of Electrodynamic Thrust Bearings. In Proceedings of the 16th International Symposium on Magnetic Bearings (ISMB), Beijing, China, 13–17 August 2018.

Received: 30 December 2018; Accepted: 29 January 2019; Published: 1 February 2019

Abstract: Electrodynamic thrust bearings (EDTBs) provide contactless rotor axial suspension through electromagnetic forces solely leaning on passive phenomena. Lately, linear state-space equations representing their quasi-static and dynamic behaviours have been developed and validated experimentally. However, to date, the exploitation of these models has been restricted to basic investigations regarding the stiffness and the rotational losses as well as qualitative stability analyses, thus not allowing us to objectively compare the intrinsic qualities of EDTBs. In this context, the present paper introduces four performance criteria directly related to the axial stiffness, the bearing energy efficiency and the minimal amount of external damping required to stabilise the thrust bearing. In addition, the stability is thoroughly examined via analytical developments based on these dynamical models. This notably leads to static and dynamic conditions that ensure the stability at a specific rotor spin speed. The resulting stable speed ranges are studied and their dependence to the axial external stiffness as well as the external non-rotating damping are analysed. Finally, a case study comparing three topologies through these performance criteria underlines that back irons fixed to the windings are not advantageous due to the significant detent force.

Keywords: performance criteria; damping; electrodynamic; energy efficiency; stability; stiffness; thrust bearing

1. Introduction

Nowadays, magnetic bearings constitute a convincing alternative to classical solutions such as ball or journal bearings by ensuring contactless guiding of rotors, thereby reducing losses and removing mechanical wear and friction. These compelling bearing can be either active or passive. The former are based on current-controlled electromagnets exerting an attractive force on a ferromagnetic rotor, whereas the latter only rely on passive phenomena.

Electrodynamic bearings (EDBs) belong to passive magnetic bearings (PMBs) as they lean on electromagnetic forces generated by the appearance of induced currents in short-circuited conductors in relative motion with respect to a magnetic field produced by permanent magnets (PMs). Although their stiffness is quite low in comparison with active magnetic bearings (AMBs), these bearings are attractive as they require neither sensors nor power and control electronics, thereby being intrinsically more reliable, compact and energy-efficient [1]. EDBs can be of two types: radial or axial bearings. The former allows guiding the radial degrees of freedom of the rotor, whereas the latter provides the axial levitation.

Numerous models describing radial EDBs in quasi-static conditions [2,3], i.e., assuming constant spin speed and eccentricity, as well as in dynamic conditions were developed [4,5]. Although they have never been defined as such, several criteria allowing us to compare these EDBs came up along with these models.

Obviously, the stiffness induced by the electrodynamic effects is of primary interest given that it directly relates to the bearing stability and eccentricity. This stiffness is an increasing function of the rotor spin speed and can be characterised through two coefficients, namely the maximal stiffness and the electrical pole of the R-L equivalent circuit [6]. Several sensitivity analyses were performed on these two coefficients, thus yielding a first insight of the geometrical [7], electrical [8] and magnetic parameters [9] that strongly influence them.

In addition to the stiffness, attention is paid to the rotational losses required to provide the levitation force. Indeed, these losses are dissipated as heat and should therefore be limited to avoid significant temperature rises as well as to increase the energy efficiency. To this end, the null-flux concept was transposed to heteropolar EDBs, allowing us to conceive new topologies whose flux linkage is null when there is no rotor eccentricity [10]. In this way, there is no induced currents and therefore no losses in this position. Similarly, the null-E concept was then developed for homopolar bearings [11]. Simultaneously, analytical formulas were derived to evaluate these rotational losses [12,13].

The dynamic behaviour of radial EDBs constitutes a major issue as these bearings are always unstable in the absence of non-rotating damping, i.e., damping that does not depend on the rotor rotation [5,14]. Considering the difficulty of adding damping in a contactless way, thus being consistent with the magnetic bearing approach, this external damping should be minimised. To this end, analytical expressions were developed on the basis of quasi-static models to determine the minimal damping required to ensure the stability at a particular spin speed [2,9,15].

Despite their promising stability properties, electrodynamic thrust bearings (EDTBs) have focused much less research efforts. A bearing energy efficiency, defined as the ratio between the electrodynamic levitation force and the corresponding power losses, has been introduced as a performance criterion, even though external stiffnesses, such as the detent one, cannot be taken into account [16]. Recently, models describing both the axial quasi-static and dynamic behaviours of EDTBs have been derived and validated experimentally, allowing us to study their stiffness and rotational losses [17–20]. By contrast, although the beneficial effect of the external damping has been theoretically demonstrated, there is still no formula allowing us to determine the additional damping required to ensure the stability. Similarly, the spin speed ranges within which the EDTB is stable can still not be determined analytically.

In this context, the present paper introduces four performance criteria related to the bearing axial stiffness, the energy efficiency and the stability, allowing us to compare objectively EDTB topologies in terms of their intrinsic qualities. Analytical expressions of these criteria are derived on the basis of the dynamic models proposed in [17,18,21], thus being suitable for a wide variety of thrust bearing. In addition, static and dynamic stabilities are analysed analytically, providing conditions that ensure that the EDTB is stable at a particular spin speed and therefore allowing us to determine the stable spin speed ranges.

The paper is structured as follows. Section 2 depicts the thrust bearing topologies under study. Following on from this, the electromechanical model, comprising the electromagnetic and the rotor mechanical models, is described in Section 3. The stability of the system is then analysed in Sections 4 and 5. Section 6 defines the four performance criteria for EDTBs. The last section is devoted to a case study analysing three topologies through these criteria.

2. Bearing Description

The thrust bearing being analysed is constituted of two independent subassemblies, namely the PM arrangements and the armature winding, in rotary motion relative to each other, as illustrated in Figure 1. Each of them can be attached either to the stator or to the rotor.

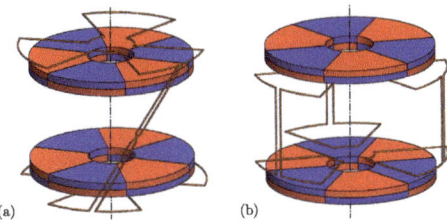

Figure 1. Bearing topologies with only one phase represented: (**a**) PMs are internal and the p coils of each set are connected in series, the two resulting sets being connected together in series; and (**b**) PMs are external and the p upper and the p lower coils are independently connected together in opposition.

The first subassembly comprises two PM arrangements, each producing an identical axial magnetic field with p pole pairs. These arrangements can:

- either be placed in repulsive or attractive mode, as represented in Figure 1; and
- either constitute the internal or external subassembly, as shown in Figure 1a,b respectively.

The armature winding comprises N identical and evenly distributed phase windings. The latter are each constituted of two identical sets of p identical and evenly distributed coils, each set being predominantly magnetically linked to one PM arrangement, and can be of two types:

- the p coils of each set are connected in series, the two resulting sets being connected together, as illustrated in Figure 1a;
- the p upper and the p lower coils are independently connected together, as represented in Figure 1b.

Besides, as illustrated in Figure 1a,b, respectively, both upper and lower sets of coils can be shifted by an angle equal to π/p or zero and can be connected either in series or in opposition. This connection is chosen on the basis of the angular shift that separates the upper and lower sets as well as the attractive or repulsive mode of the PM arrangements so as to ensure that the flux linked by the armature winding is null when the rotor is axially centred with respect to the stator, thereby respecting the null-flux principle.

3. Electromechanical Model

Under the assumption of small rotor axial, radial and angular displacements and neglecting the inductance coefficient variations with these displacements, the axial dynamics of the system constituted of the rotor and the ETDB is decoupled from the radial and angular ones [22]. Assuming in addition that the rotor spin speed varies slowly compared to the axial dynamics, the latter can be described through a linear state-space representation as extensively derived in [18]:

$$\begin{bmatrix} \ddot{z} \\ \dot{z} \\ \dot{F} \\ \left(\dfrac{\dot{T}}{z}\right) \end{bmatrix} = \mathbf{A} \begin{bmatrix} \dot{z} \\ z \\ F \\ \left(\dfrac{T}{z}\right) \end{bmatrix} + \mathbf{B} \cdot F_e, \tag{1}$$

with:

$$\mathbf{A} = \begin{bmatrix} -\dfrac{C}{M} & -\dfrac{k_e}{M} & \dfrac{1}{M} & 0 \\ 1 & 0 & 0 & 0 \\ -\dfrac{K_\Phi^2 N}{2L_c} & 0 & -\dfrac{R}{L_c} & \omega \\ 0 & -\omega p^2 \left(\dfrac{K_\Phi^2 N}{2L_c}\right) & -\omega p^2 & -\dfrac{R}{L_c} \end{bmatrix}, \quad (2)$$

$$\mathbf{B} = \dfrac{1}{M}\begin{bmatrix} 1 & 0 & 0 & 0 \end{bmatrix}^T, \quad (3)$$

where z and $\dot{z}$ are, respectively, the rotor axial position and velocity; F and T are, respectively, the electrodynamic force and torque; F_e is the external axial force acting on the rotor; C is the external non-rotating damping; M is the rotor mass; R is the phase winding resistance; L_c is the cyclic inductance, thus taking into account the self and mutual inductance coefficients of the N phases constituting the armature winding; K_Φ is the proportionality factor between the amplitude of the flux linked by the phase windings due to the PMs and the axial position; and k_e is the external axial stiffness. The latter could, for example, arise from detent effects or be related to the axial stiffness induced by centring PM bearings added to the system so as to ensure the rotor radial and angular guidance. Hence, this stiffness is generally negative, as it is assumed hereafter.

Assuming quasi-static conditions, i.e., $\dot{z} = 0$, the axial electrodynamic stiffness $k(\omega)$ as well as the associated braking torque $T(\omega)$ can be retrieved from this dynamic model, yielding [18]:

$$k(\omega) = -\dfrac{F(\omega)}{z} = \dfrac{K_\Phi^2 N}{2L_c}\dfrac{\omega^2}{\omega^2 + \left(\dfrac{1}{p}\dfrac{R}{L_c}\right)^2}$$
$$T(\omega) = -z^2\dfrac{K_\Phi^2 N}{2L_c}\dfrac{R}{L_c}\dfrac{\omega}{\omega^2 + \left(\dfrac{1}{p}\dfrac{R}{L_c}\right)^2} \quad (4)$$

As depicted in Figure 2, illustrating the evolution of the stiffness, the latter increases with the spin speed and can be characterised through two coefficients, namely the rotor spin speed $\omega_e = R/(pL_c)$ related to the electrical pole and the asymptotic stiffness k_∞, defined as:

$$k_\infty = \dfrac{K_\Phi^2 N}{2L_c} \quad (5)$$

The latter therefore corresponds to the maximal axial stiffness that can be generated by the EDTB. Let us point out that this stiffness appears explicitly in the state matrix $\mathbf{A}$, given in Equation (2). On the contrary, as shown in Figure 2, the braking torque $T(\omega)$ reaches its maximal value when the speed is equal to ω_e and then decreases asymptotically to zero.

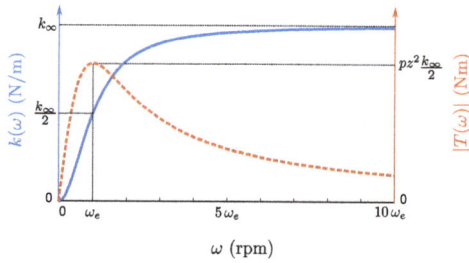

Figure 2. Evolution of the electrodynamic stiffness k (solid line) and braking torque T (dashed line) with the spin speed ω.

4. Stability Analysis

The behaviour of EDTBs is strongly dependent on the rotor spin speed and so is their stability. Hereinafter, general considerations about the stability of an EDTB coupled to the rotor are first derived. On this basis, the static and dynamic stability are then analysed, leading to conditions ensuring a stable behaviour at a specific rotor spin speed.

The following developments can be greatly simplified by considering the electrical pole as being much greater than the maximal natural frequency of the equivalent spring–mass system constituted of the rotor and the EDTB:

$$\frac{R}{L_c} \gg \sqrt{\frac{k_\infty}{M}}. \tag{6}$$

In this way, the electrical phenomena are much faster than the mechanical ones and thus do not have a significant impact on the rotor axial dynamics. Observing that the electromechanical model, given in Equation (1), depends on the stiffness as well as the rotor mass and not their square roots, Equation (6) can be expressed in a more convenient manner as:

$$\left(\frac{R}{L_c}\right)^2 \gg \frac{k_\infty}{M}. \tag{7}$$

To the authors' best knowledge, the latter hypothesis is verified in the vast majority of the experimental and numerical studies of EDTBs, including the case study in Section 7. In addition, let us assume a priori that the external damping satisfies:

$$2\left(\frac{C}{M}\right) \ll \frac{R}{L_c}, \tag{8}$$

This assumption is verified a posteriori in Section 4.3.

4.1. General Considerations

The model in Equation (1) being linear, the stability analysis can be performed through the study of the real part of the four eigenvalues of the state matrix **A** as a function of the spin speed. To this end, the characteristic polynomial can be easily derived, yielding:

$$\begin{aligned}P(s) =\ & s^4 + s^3\left(\frac{C}{M} + \frac{2R}{L_c}\right) + s^2\left(\frac{C}{M}\frac{2R}{L_c} + \frac{R^2}{L_c^2} + \omega^2 p^2 + \frac{k_\infty + k_e}{M}\right) \\ & + s\left(\frac{C}{M}\frac{R^2}{L_c^2} + \omega^2 p^2 \frac{C}{M} + \frac{R}{L_c}\frac{k_\infty + 2k_e}{M}\right) + \omega^2 p^2 \frac{k_\infty + k_e}{M} + \frac{R^2}{L_c^2}\frac{k_e}{M}. \end{aligned} \tag{9}$$

Under the hypothesis expressed in Equation (7) and assuming Equation (8) as verified, the polynomial in Equation (9) can be simplified as follows:

$$P(s) = s^4 + s^3\frac{2R}{L_c} + s^2\left(\frac{R^2}{L_c^2} + \omega^2 p^2\right) + s\left(\frac{C}{M}\frac{R^2}{L_c^2} + \omega^2 p^2 \frac{C}{M} + \frac{R}{L_c}\frac{k_\infty + 2k_e}{M}\right) + \omega^2 p^2 \frac{k_\infty + k_e}{M} + \frac{R^2}{L_c^2}\frac{k_e}{M}. \tag{10}$$

The root locus of the four eigenvalues can thus be obtained by finding the roots of Equation (10) for different spin speeds. However, when it comes to stability analyses, only the speeds at which the eigenvalues cross the imaginary axis are relevant as they define the spin speed ranges within which the bearing is stable. Figure 3a,b illustrates, respectively, the impact of the external damping and stiffness on the root locus. Only the two relevant eigenvalues, related to the mechanical behaviour, are represented, the remaining two, related to the electrical behaviour, being located far in the left half plane. The additional damping allows us to shift the complex conjugates parts of the root locus to the left by an amount equal to $C/(2M)$, whereas the external stiffness strongly modifies their shape.

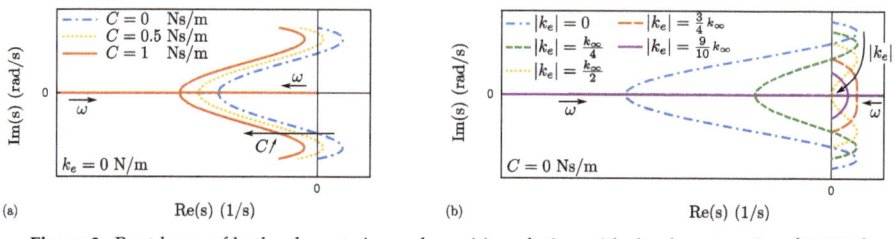

Figure 3. Root locus of both relevant eigenvalues: (a) evolution with the damping $C = \{0, 0.5, 1\}$ Ns/m for $k_e = 0$ N/m; and (b) evolution with the external stiffness $|k_e| = \{0, \frac{1}{4}, \frac{1}{2}, \frac{3}{4}, \frac{9}{10}\} k_\infty$ N/m for $C = 0$ Ns/m.

As a result, there are at most three spin speeds, ω_1, ω_2 and ω_3, defined in Figure 4, corresponding to intersections with the imaginary axis. More precisely, as shown in Figure 3, when the external damping approaches zero, the spin speed ω_3 tends to infinity and therefore no longer exists. By contrast, increasing the damping allows us to move the spin speeds ω_2 and ω_3 towards each other until they are equal, when the damping reaches a specific value, denoted by C_m hereinafter. Beyond the latter damping, these two speeds do not exist anymore. Besides, as illustrated in Figure 3b, the presence of the speed ω_2 strongly depends on the external stiffness.

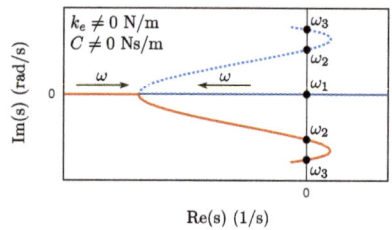

Figure 4. Root locus: Spin speeds corresponding to intersections with the imaginary axis.

For determining these speeds, let us assume that $s = jh$, implying that the eigenvalue lies on the imaginary axis. In this case, Equation (10) can be separated into real and imaginary parts as follows:

$$\begin{cases} 0 = h^4 + \omega^2 p^2 \dfrac{k_\infty + k_e}{M} + \dfrac{R^2}{L_c^2} \dfrac{k_e}{M} - h^2 \left(\dfrac{R^2}{L_c^2} + \omega^2 p^2 \right) & (11) \\ 0 = h \left(\dfrac{C}{M} \dfrac{R^2}{L_c^2} + \omega^2 p^2 \dfrac{C}{M} + \dfrac{R}{L_c} \dfrac{k_\infty + 2k_e}{M} \right) - h^3 \dfrac{2R}{L_c} & (12) \end{cases}$$

Solving Equation (12) for h yields three solutions. As demonstrated hereinafter, one solution is related to a static instability, whereas the other two are linked to a dynamic one.

4.2. Static Stability

The trivial solution of Equation (12), i.e., $h = 0$, corresponds to the first intersection of the eigenvalues with the imaginary axis. Substituting this solution into Equation (11) and isolating ω leads to:

$$\omega_1 = \frac{1}{p} \frac{R}{L_c} \sqrt{-\frac{k_e}{k_e + k_\infty}}. \tag{13}$$

This corresponds to the spin speed at which the stiffness induced by the electrodynamic effects exactly compensates for the external stiffness, i.e., $k(\omega_1) = |k_e|$, as can be verified through Equation (4). Below this specific spin speed, the thrust bearing suffers from an instability as the external stiffness, whose effect is destabilising due to its negative value, is larger than the electrodynamic one.

This instability can be qualified as static as it does not depend on the damping. The static stability condition can thus be stated as:

$$k(\omega) \geq |k_e| \iff \omega \geq \omega_1 \tag{14}$$

Two limiting cases can be studied. On the one hand, when there is no external stiffness, the speed ω_1 is equal to zero and the static stability condition does not introduce any restriction on the rotor spin speed. On the other hand, when the external stiffness is larger, in absolute value, than the maximal electrodynamic stiffness, i.e., $|k_e| > k_\infty$, the speed ω_1 tends to infinity and the bearing is unstable regardless of the rotor spin speed.

4.3. Dynamic Stability

Both remaining solutions of Equation (12) are linked to a dynamic instability as they depend on the damping. They can be calculated as follows:

$$h = \pm \sqrt{\frac{1}{2} \frac{C}{M} \frac{R}{L_c} + \omega^2 p^2 \frac{1}{2} \frac{C}{M} \frac{L_c}{R} + \frac{k_\infty + 2k_e}{2M}}. \tag{15}$$

Substituting Equation (15) into Equation (11) and multiplying by $4R^2/L_c^2$ yields:

$$\omega^4 f_1 + \omega^2 f_2 + f_3 = 0, \tag{16}$$

where:

$$f_1 = p^4 \frac{C}{M} \left[\frac{C}{M} - 2\frac{R}{L_c} \right]$$

$$f_2 = 2p^2 \frac{R}{L_c} \left[\frac{R}{L_c} \frac{k_\infty}{M} + \frac{R}{L_c} \left(\frac{C}{M} \right)^2 + \frac{k_\infty + 2k_e}{M} \frac{C}{M} - 2 \left(\frac{R}{L_c} \right)^2 \frac{C}{M} \right] \tag{17}$$

$$f_3 = \left(\frac{R}{L_c} \right)^2 \left[\left(\frac{R}{L_c} \frac{C}{M} \right)^2 + \left(\frac{k_\infty + 2k_e}{M} \right)^2 - 2 \left(\frac{R}{L_c} \right)^3 \frac{C}{M} - 2 \left(\frac{R}{L_c} \right)^2 \frac{k_\infty}{M} + 2 \frac{R}{L_c} \frac{C}{M} \frac{k_\infty + 2k_e}{M} \right].$$

The polynomial in Equation (16) has at most two positive roots, thereby confirming that both eigenvalues related to the electrical behaviour never cross the imaginary axis. Under the hypothesis expressed in Equation (7) and still assuming that the damping satisfies Equation (8), the coefficients in Equation (17) can be greatly simplified, leading to:

$$f_1 = -2p^4 \frac{C}{M} \frac{R}{L_c} \tag{18}$$

$$f_2 = 2p^2 \left(\frac{R}{L_c} \right)^2 \left[\frac{k_\infty}{M} - 2\frac{R}{L_c} \frac{C}{M} \right] \tag{19}$$

$$f_3 = -2 \left(\frac{R}{L_c} \right)^4 \left[\frac{R}{L_c} \frac{C}{M} + \frac{k_\infty}{M} \right]. \tag{20}$$

Solving Equation (16) with these reduced coefficients allows us to determine both spin speeds ω_2 and ω_3 at which the relevant eigenvalues cross the imaginary axis, as shown in Figure 4:

$$\begin{cases} \omega_{2,3}|_{C \neq 0} = \frac{1}{p} \sqrt{\frac{1}{2} \frac{R}{L_c} \frac{M}{C} \left[\frac{k_\infty}{M} - 2\frac{C}{M} \frac{R}{L_c} \mp \sqrt{\Delta} \right]} \\ \Delta = \left(\frac{k_\infty}{M} \right)^2 - 8 \frac{R}{L_c} \frac{C}{M} \frac{k_\infty}{M} \end{cases}. \tag{21}$$

The value of these speeds is independent from the external stiffness k_e, signifying that the intersections of the eigenvalues with the imaginary axis occur at the same spin speeds even when the shape of the root locus is modified by this stiffness, as shown in Figure 3b. By contrast, as mentioned in Section 4.1, the existence of these intersections strongly depends on the external damping and stiffness.

Figure 5 shows the evolution of both speeds ω_2 and ω_3 with the external damping. As expected, when the latter is equal to zero, the speed ω_3 tends to infinity and therefore no longer exists, whereas the speed ω_2 can be easily calculated by observing that the coefficient f_1 in Equation (18) is equal to zero, implying that Equation (16) has only one positive solution:

$$\omega_2|_{C=0} = \frac{1}{p}\frac{R}{L_c} = \omega_e. \tag{22}$$

This speed thus corresponds to spin speed ω_e related to the electrical pole. Let us point out that spin speeds smaller than this particular speed can never lie on the imaginary axis and are therefore stable, from a dynamic point of view, regardless of the damping. As stated in Section 4.1, adding external damping enables moving the speeds ω_2 and ω_3 towards each other until they intersect, when the damping reaches C_m. Cancelling the coefficient Δ in Equation (21) allows us to determine both the damping C_m such that these two speeds are equal and the corresponding speed, denoted by ω_m:

$$\begin{cases} C_m = \dfrac{k_\infty}{8}\dfrac{L_c}{R} = \dfrac{K_\Phi^2 N}{16R} & (23) \\ \omega_m = \dfrac{\sqrt{3}}{p}\dfrac{R}{L_c} & (24) \end{cases}$$

Below this damping, the speeds ω_2 and ω_3 are distinct and the EDTB is unstable, from a dynamic point of view, when the spin speed belongs to the interval $[\omega_2\,;\,\omega_3]$, as shown in Figure 4. By contrast, when the damping is larger than C_m, the eigenvalues only cross the imaginary axis at the speed ω_1 and the EDTB is stable beyond the latter speed. Consequently, unlike their static counterparts, dynamic instabilities can be removed through additional non-rotating damping.

Finally, substituting the maximal damping given in Equation (23) into Equation (8) and considering that the assumption in Equation (7) is verified allows us to validate the relation in Equation (8) a posteriori, highlighting that the latter is not, as such, a hypothesis.

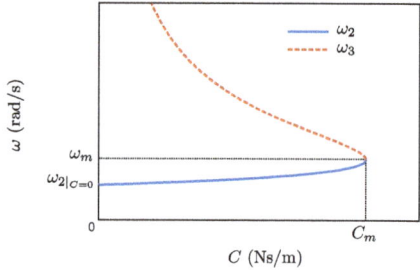

Figure 5. Evolution of the speeds ω_2 and ω_3 with the external damping C.

4.4. Stability Conditions

In summary, the stability can be analysed on the basis of:

- the speed ω_1 related to the static instability, given in Equation (13);
- the speeds ω_2 and ω_3, related to the dynamic instability, as a function of the damping, defined in Equation (21); and
- the external damping C added to the system.

More precisely, when the maximal electrodynamic stiffness is larger than the external one, i.e., $k_\infty > |k_e|$, the stability is ensured at the spin speed ω provided that:

$$\begin{cases} \omega \leq \omega_2(C) \text{ or } \omega \geq \omega_3(C) & \text{if } C \in [0\,;\,C_m] \\ \omega \geq \omega_1 \end{cases} \quad (25)$$

Finally, let us point out that the state-space representations in [17,21] yield an identical characteristic polynomial to Equation (9), thus widening the scope of the previous developments to these models.

5. Stable Speed Analysis

Assuming that the eight parameters describing the dynamic behaviour of the system are identified, the speed ranges within which the EDTB is stable can be easily determined through the conditions defined in Equation (25). However, let us go one step further by analysing three different cases, depending on the relative importance of the external stiffness in comparison to the electrodynamic one.

5.1. $|k_e| \in [0\,;\,\frac{k_\infty}{2}]$

Let us first consider that the external stiffness belongs, in absolute value, to the interval $[0\,;\,\frac{k_\infty}{2}]$. In this case, the spin speed ω_1 lies between 0 and $\omega_2|_{C=0}$. Figure 6a,b represents, respectively, the different curves involved in the stability conditions and the corresponding root locus. As shown in Figure 6a, the EDTB is stable when the spin speed belongs to $[\omega_1\,;\,\omega_2]$ or $[\omega_3\,;\,\infty[$. By contrast, when the external damping is equal to zero, the intersection linked to spin speed ω_3 does not exist and the bearing is stable only between ω_1 and $\omega_2|_{C=0}$. Finally, when the damping is larger than C_m, both speeds ω_2 and ω_3 no longer exist and the stability range is enlarged to the interval $[\omega_1\,;\,\infty[$.

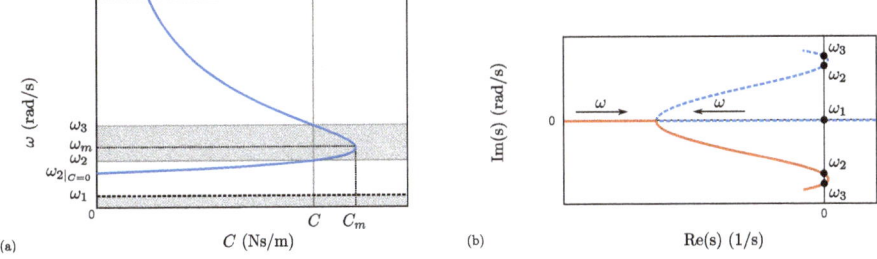

Figure 6. Stability analysis for $|k_e| \in [0\,;\,\frac{k_\infty}{2}]$: (**a**) evolution of the spin speeds ω_1, ω_2 and ω_3 with the external damping, yielding the stable spin speed ranges; and (**b**) the corresponding root locus.

5.2. $|k_e| \in [\frac{k_\infty}{2}\,;\,\frac{3k_\infty}{4}]$

Considering then the case with the external stiffness belonging to the interval $[\frac{k_\infty}{2}\,;\,\frac{3k_\infty}{4}]$, the speed ω_1 can vary from $\omega_2|_{C=0}$ to ω_m. Figure 7a,b represents, respectively, the different curves involved in the stability conditions and the corresponding root locus. In this case, the EDTB is stable when the spin speed belongs to $[\omega_1\,;\,\omega_2]$ or $[\omega_3\,;\,\infty[$ provided that the damping C is larger than the damping C_1 related to ω_1, as shown in Figure 7a. The latter damping can be easily calculated by inverting Equation (21) and evaluating the resulting function at the speed ω_1, yielding:

$$C_1 = -k_\infty \frac{L_c}{R} \frac{\left[1 + \frac{k_e}{k_\infty + k_e}\right]}{\left[1 - \frac{k_e}{k_\infty + k_e}\right]^2}. \quad (26)$$

By contrast, when the additional damping is smaller than C_1, the stability range is limited to the interval $[\omega_3\,;\,\infty[$ given that the speed ω_2 no longer corresponds to an intersection with the imaginary axis, the latter speed being smaller than ω_1. Let us point out that, when there is no external damping, the system suffers from a dynamic instability for speeds larger than ω_1 and is therefore unconditionally unstable. Finally, when the damping is larger than C_m, the stable spin speed range is $[\omega_1\,;\,\infty[$.

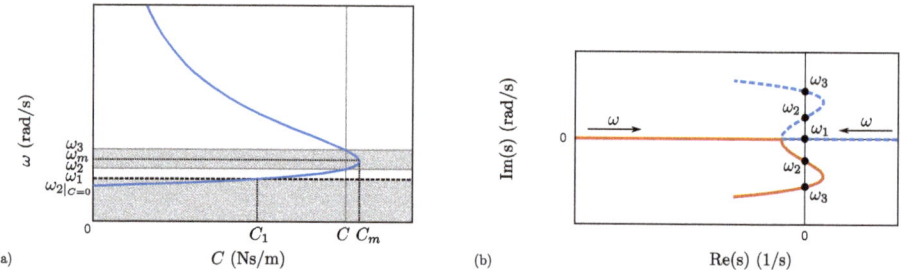

Figure 7. Stability analysis for $|k_e| \in [\frac{k_\infty}{2}\,;\,\frac{3k_\infty}{4}]$: (**a**) evolution of the spin speeds ω_1, ω_2 and ω_3 with the external damping, yielding the stable spin speed ranges; and (**b**) the corresponding root locus.

5.3. $|k_e| \in [\frac{3k_\infty}{4}\,;\,k_\infty]$

Let us now consider the case with an external stiffness belonging to the interval $[\frac{3k_\infty}{4}\,;\,k_\infty]$, implying that the speed ω_1 is larger than ω_m. Figure 8a,b represents, respectively, the different curves involved in the stability conditions and the corresponding root locus. In this last case, the stability range corresponds to the interval is $[\omega_3\,;\,\infty[$ provided that the damping C is smaller than C_1, as shown in Figure 8a. Otherwise, the stable speed range is given by $[\omega_1\,;\,\infty[$. Let us point out that adding an amount of external damping larger than C_1 brings no benefits in terms of stability. Finally, when the damping is equal to zero, the bearing is unconditionally unstable.

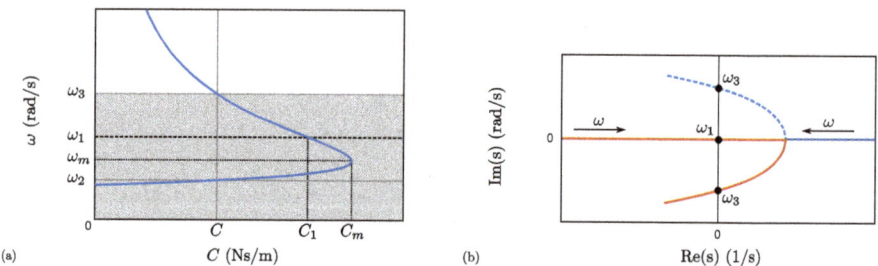

Figure 8. Stability analysis for $|k_e| \in [\frac{3k_\infty}{4}\,;\,k_\infty]$: (**a**) evolution of the spin speeds ω_1, ω_2 and ω_3 with the external damping, yielding the stable spin speed ranges; and (**b**) the corresponding root locus.

5.4. Summary

Table 1 summarises the intervals within which the axial dynamics of the system constituted of the EDTB coupled to the rotor is stable, depending on the external damping and stiffness.

Table 1. Stable speed ranges.

$\|k_e\|$	C	Interval
	0	$[\omega_1; \omega_2]$
$[0; \frac{k_\infty}{2}]$	$]0; C_m]$	$[\omega_1; \omega_2] \cup [\omega_3; \infty[$
	$]C_m; \infty[$	$[\omega_1; \infty[$
	0	$\emptyset$
	$]0; C_1]$	$[\omega_3; \infty[$
$[\frac{k_\infty}{2}; \frac{3k_\infty}{4}]$	$]C_1; C_m]$	$[\omega_1; \omega_2] \cup [\omega_3; \infty[$
	$]C_m; \infty[$	$[\omega_1; \infty[$
	0	$\emptyset$
$[\frac{3k_\infty}{4}; k_\infty]$	$]0; C_1]$	$[\omega_3; \infty[$
	$]C_1; \infty[$	$[\omega_1; \infty[$

6. Performance Criteria

The stiffness, the losses and the stability are of primary interest when analysing a bearing. On this basis, four criteria can be derived to evaluate the intrinsic qualities of EDTB topologies, thus allowing us to compare them objectively. These criteria are independent from the rotor spin speed as well as its axial displacement.

6.1. Total Stiffness

In quasi-static conditions, the total stiffness $k_t(\omega)$, comprising both electrodynamic and external effects, can be expressed as follows:

$$k_t(\omega) = -\frac{F_t(z,\omega)}{z} = k(\omega) + k_e. \tag{27}$$

As stated above, the static stability of the system as well as the rotor axial position and dynamics are directly related to this stiffness. The maximal total stiffness $k_{t,\infty}$ therefore constitutes a first performance criterion to be maximised:

$$k_{t,\infty} = k_\infty + k_e = \frac{K_\Phi^2 N}{2L_c} + k_e, \tag{28}$$

Further noting that, for fixed maximal stiffness $k_{t,\infty}$ and speed ω, decreasing the spin speed ω_e corresponding to the electrical pole R/L_c allows us to increase the stiffness, the latter speed constitutes a second criterion to be minimised:

$$\omega_e = \frac{1}{p}\frac{R}{L_c}. \tag{29}$$

6.2. Stability Margin

As stated above, adding non-rotating damping allows us to enlarge the range within which the system is stable. However, to be coherent with the magnetic bearing approach, the external damping should be contactless. Considering the potential difficulty of producing the latter, the damping C_s required to stabilise the thrust bearing regardless of the spin speed should be minimised. This is all the more true observing that maximising the stiffness and thus minimising the speed corresponding to the electrical pole reduces the stable speed range when there is no external damping. As mentioned in Section 5, this damping C_s depends on the relative importance of the external stiffness in comparison to the maximal electrodynamic one:

$$C_s = \begin{cases} C_m & \text{if } |k_e| \in \left[0; \frac{3k_\infty}{4}\right] \\ C_1 & \text{if } |k_e| \in \left[\frac{3k_\infty}{4}; k_\infty\right] \end{cases}, \tag{30}$$

where C_m and C_1 can be respectively calculated through Equations (23) and (26).

6.3. Energy Efficiency Coefficient

In addition to the restoring force, the thrust bearing produces an electrodynamic braking torque, therefore contributing to decrease the rotor spin speed. The power P related to this braking torque is entirely dissipated in the winding resistances in the form of Joule losses, leading to a rise in temperature and thus being potentially detrimental to the functioning of the bearing. In quasi-static conditions, these rotational losses can be calculated as:

$$P(\omega) = |\omega\, T(\omega)| = z^2 k_\infty \frac{R}{L_c} \frac{\omega^2}{\omega^2 + \left(\frac{1}{p}\frac{R}{L_c}\right)^2}. \tag{31}$$

The bearing purpose is to provide the largest axial levitation force F_t, whereas the associated rotational losses P have to be minimised. This amounts to maximising the following ratio:

$$\frac{F_t}{\sqrt{P}} = \sqrt{\frac{(k_\infty + k_e)^2}{k_\infty} \frac{L_c}{R} \frac{[(p\omega)^2 - (p\omega_1)^2]^2}{(p\omega)^2 \left[(p\omega)^2 + \left(\frac{R}{L_c}\right)^2\right]}}. \tag{32}$$

This ratio therefore only exists for rotor spin speeds larger than ω_1, increasing from zero up to reach its asymptotic value denoted by K_p:

$$K_p = \sqrt{\frac{(k_\infty + k_e)^2}{k_\infty} \frac{L_c}{R}}. \tag{33}$$

The energy efficiency coefficient K_p thus constitutes a fourth performance criterion to be maximised. Lastly, in the absence of external stiffness, Equation (33) reduces to:

$$K_p\big|_{k_e=0} = \sqrt{\frac{L_c}{R} k_\infty}. \tag{34}$$

The latter coefficient is proportional to the square root of the external damping C_m required to stabilise the bearing regardless of the spin speed, given in Equation (23). However, the energy efficiency has to be maximised, whereas the additional damping has to be minimised. A trade-off between these two criteria must therefore be considered, depending in particular on the application requirements as regards losses and spin speed.

6.4. Summary

Table 2 summarises the four performance criteria that have been derived hereinbefore.

Table 2. Performance criteria.

	Criterion	Expression
Total stiffness	max $k_{t,\infty}$	Equation (28)
	min ω_e	Equation (29)
Required damping	min C_s	Equation (30)
Energy efficiency coefficient	max K_p	Equation (33)

7. Case Study

The case study was performed on the three EDTBs illustrated in Figure 9. The first corresponds to a topology with a merged armature winding as internal subassembly and is denominated Topology 1. The second bearing, denominated Topology 2, corresponds to the topology with two distinct PM arrangements as internal subassembly and the armature winding consisting of two sets of p coils connected in series, the two resulting sets being themselves connected in opposition. The last one, denominated Topology 3, is identical to the second but includes in addition back irons on which the sets of coils are placed. In each of these three topologies, the PM arrangements comprise ferromagnetic yokes, the remanent magnetisation is 1.42 T and the number p of pole pairs is two. The armature winding comprises three phases ($N = 3$) and the conductor density, defined as the number of conductors per unit of coil section, is 4 per square millimetre. The rotor includes the armature winding and its mass was set to 1 kg. Lastly, the overall dimensions of the three topologies, given in Table 3, are identical and so is their PM volume.

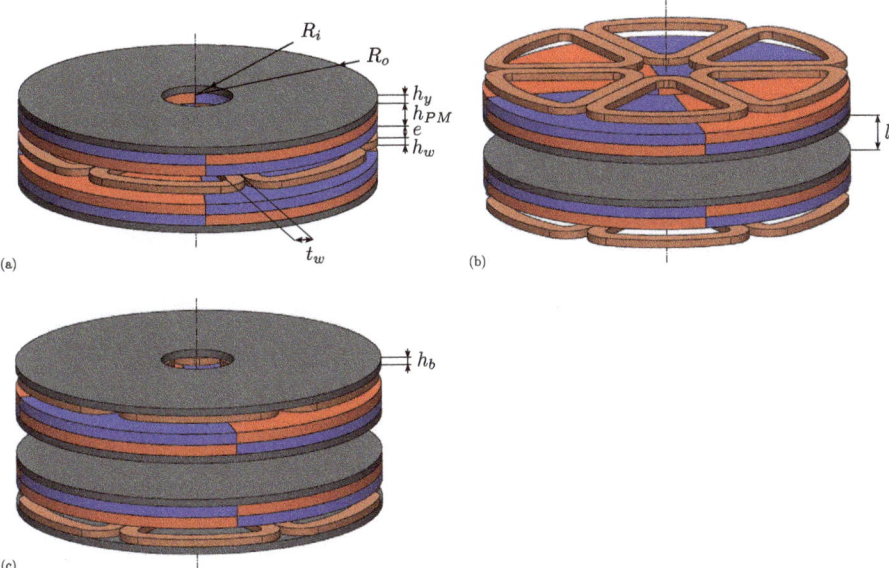

Figure 9. Study case: bearing topologies: (**a**) Topology 1; (**b**) Topology 2; and (**c**) Topology 3.

Table 3. Study case: bearing dimensions (mm).

R_i	R_o	h_y	h_{PM}	e	t_w	l	h_b	h_w
10	50	2	3	3	5	10	2	parameter

7.1. Parametric Analysis

For each topology, a parametric analysis of the four performance criteria defined above was performed with respect to the winding thickness h_w. To this end, the model parameters were identified for all configurations through static finite element simulations by applying the methods detailed in [18]. As illustrated in Figure 10a, the square of the ratio between the natural frequency of the equivalent spring–mass system and the electrical pole stayed below 7%, therefore validating the assumption in Equation (7) as well as the resulting developments with regard to the stability analyses and the external damping required to stabilise the bearing.

Figure 10b shows the evolution of the maximal total stiffness with the winding thickness. Topology 1 reached its maximum, namely 25.5 N/mm, when the thickness was equal to 10 mm, whereas Topology 2 had a peak value of 23.5 N/mm for a thickness of 2 mm. Furthermore, below a thickness of about 6.2 mm, represented by a dotted line, the total stiffness of the third topology was negative, meaning that the detent force due to the interaction between the PMs and the back irons was larger than the electrodynamic one and thus leading to a static instability regardless of the speed. Above this particular thickness, the maximal stiffness increased until it joined the curve related to Topology 2 without ever exceeding the latter. The presence of the back irons is therefore clearly not advantageous as regards the axial stiffness.

Figure 10c shows the evolution of the spin speed ω_e corresponding to the electrical pole with the winding thickness. Regardless of the latter, Topology 1 showed smaller speeds ω_e than Topology 2, signifying that the stiffness reached its maximum at lower speeds. However, the discrepancy between these two topologies decreased with the thickness. As regards Topology 3, as soon as the total stiffness became positive, the electrical pole remained smaller than the one related to Topology 2 given that the back irons allows us to increase the cyclic inductance L_c while maintaining the resistance R unchanged.

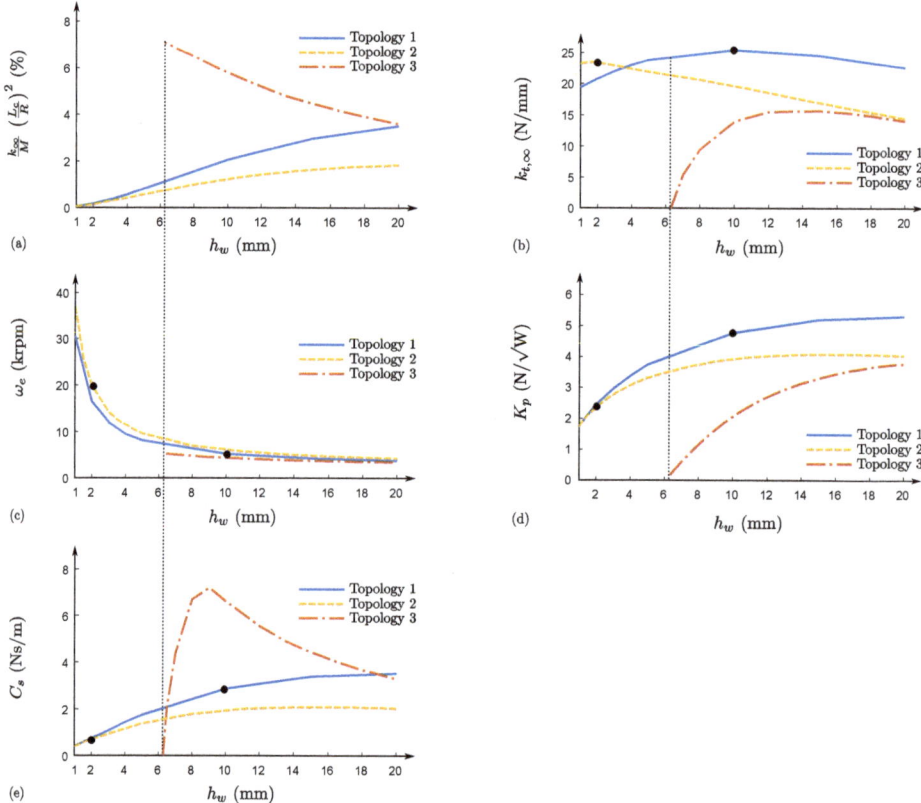

Figure 10. Evolution of the performance criteria with the winding thickness for the three topologies: (**a**) hypothesis validation; (**b**) maximal total stiffness; (**c**) spin speed related to the electrical pole; (**d**) energy efficiency coefficient; and (**e**) stability margin.

Figure 10d shows the evolution of the energy efficiency coefficient K_p with the winding thickness. As regards this criterion, Topologies 1 and 2 were rather close for small thicknesses. However, the former always outclassed the latter and the gap widened with the winding thickness. In comparison with these two topologies, the efficiency of Topology 3 remained quite low due to the negative contribution of the axial detent force.

Figure 10e shows the evolution of the damping required to stabilise the bearing at high speeds with the winding thickness. As mentioned in Section 6.3, the damping related to the Topologies 1 and 2 presented an identical shape to the curves linked to the energy efficiency, given that the latter is proportional to the square root of the required damping in the absence of external stiffness.

More precisely, it remained limited to relatively small values, namely no more than 2.0 and 3.5 Ns/m, respectively. By contrast, Topology 3 required slightly larger damping with up to the double, i.e., 7.2 Ns/m.

In summary, Topology 1 is attractive for a winding thickness close to 10 mm as the stiffness $k_{t,\infty}$ is maximal, whereas the spin speed related to the electrical pole is rather low, namely 5200 rpm. Besides, the energy efficiency coefficient is important and the required damping, being equal to 2.9 Ns/m, can be considered as reasonable in light of the values reported in the literature [23]. Topology 2 with a winding thickness equal to 2 mm yields an almost equivalent maximal stiffness, although the electrical pole is about four times larger. The required damping is thus smaller for this topology, being equal to 0.71 Ns/m, and therefore easier to produce. However, it also means that the energy efficiency is reduced by a factor about 2. Let us point out that, without considering the distance l between both parts of Topology 2, the volume occupied by both topologies is nearly identical. Only Topologies 1 and 2 with a winding thickness equal to 10 and 2 mm, respectively, were further considered.

7.2. Rotational Losses

Assuming a constant external force F_e, the resulting axial displacement can be determined through Equation (1) as well as Equation (4) and then substituted into Equation (31) giving the rotational losses, yielding:

$$P(F_e, \omega) = \underbrace{\frac{F_e^2 k_\infty}{(k_\infty + k_e)^2}}_{P_{F_e,\infty}} \frac{R}{L_c} \frac{(p\omega)^2 \left[(p\omega)^2 + \left(\frac{R}{L_c}\right)^2\right]}{\left[(p\omega)^2 + \left(\frac{R}{L_c}\right)^2 \frac{k_e}{(k_\infty + k_e)}\right]^2}. \tag{35}$$

Considering the rotor weight as external load, namely approximately 10 N, the minimal rotational losses $P_{F_e,\infty}$, given in Equation (35), were, respectively, equal to 4.4 and 17.8 W for Topologies 1 and 2. Topology 1 therefore dissipated about four times less power for an identical load. Indeed, in the absence of external stiffness, these losses were inversely proportional to the damping C_m required to stabilise the thrust bearing regardless of the spin speed.

7.3. Stiffness Analysis

We studied the evolution of the stiffness with the rotor spin speed for Topologies 1 and 2. Figure 11a,b represents, respectively, while taking into account the stability conditions for each spin speed and amount of external damping, the maximal stiffness among both topologies and the corresponding topology. The solid and dashed lines illustrate, respectively, the stability boundary related to Topologies 1 and 2, the latter being defined as the evolution with the damping of the speeds $\omega_{2,3}$ given in Equation (21). In this way, below each curve, the corresponding topology suffers from a dynamic instability, also implying that both are unstable in the white zone.

Regardless of the spin speed, Topology 1 provided a higher stiffness and reached its maximal stiffness k_∞ for lower speeds than Topology 2 given that the electrical pole was smaller. By contrast, in the absence of additional damping, Topology 1 was also unstable for a smaller speed. Indeed, as stated in Section 6.1, minimising the spin speed ω_e related to the electrical pole amounts to reducing the stable speed range when there is no external damping. Therefore, between about 5000 and 20,000 rpm, namely the spin speeds related to the electrical poles of both topologies, Topology 2 offers the major advantage of not requiring external damping to ensure the axial stable levitation of the rotor. This brief analysis shows that the rotor spin speed can still strongly influence the bearing selection according to the application specifications.

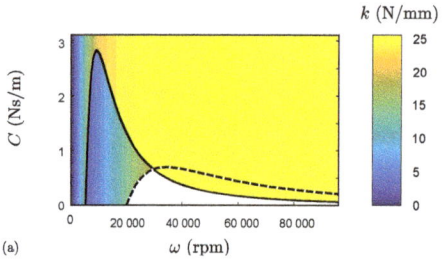 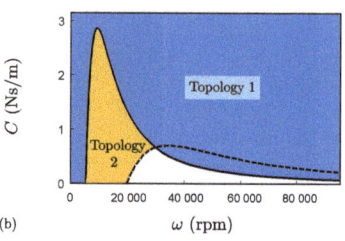

Figure 11. Comparison of the stiffness of Topologies 1 and 2 while taking into account their stability boundaries (solid and dashed lines respectively): (**a**) maximal stiffness with the spin speed; and (**b**) the corresponding topology.

8. Conclusions

This paper presents four criteria allowing us to compare objectively various electrodynamic thrust bearing topologies based on their intrinsic qualities and therefore to determine the most appropriate.

On the basis of the recent linear state-space representations describing the axial dynamics of EDTBs, an analytical static and dynamic stability analysis is performed through the calculation of the eigenvalues of the state matrix. The impact of the external damping and stiffness is studied through a root locus as a function of the rotor spin speed, highlighting that the former allows us to move the eigenvalues to the left, thus improving the stability, whereas the latter modifies their shape. Besides, the spin speeds corresponding to intersections with the imaginary axis are calculated, therefore defining the ranges within which the thrust bearing is stable. In the absence of additional damping and external stiffness, the thrust bearing is stable up to the spin speed related to the electrical pole.

When it comes to comparing magnetic bearings, the maximal eccentricity, the losses and the stability are of primary interest. As a result, the following four performance criteria are defined: (i) the maximal total stiffness; (ii) the spin speed corresponding to the electrical pole; (iii) the levitation energy efficiency, defined as the ratio between the thrust force and the corresponding rotational losses; and (iiii) the damping required to stabilise the bearing regardless of the rotor spin speed. Three different thrust bearing topologies, studied in the framework of a case study, are finally compared on the basis of these criteria, notably highlighting that the addition of back irons behind the sets of coils has no beneficial effect as regards axial dynamics due to the important detent stiffness.

Author Contributions: Conceptualization, J.V.V.; Funding acquisition, J.V.V.; Investigation, J.V.V.; Methodology, J.V.V.; Supervision, B.D.; Writing—original draft, J.V.V.; Writing—review & editing, V.K. and B.D.

Funding: J. Van Verdeghem is a FRIA Grant Holder of the Fonds de la Recherche Scientifique-FNRS, Belgium.

Conflicts of Interest: The authors declare no conflict of interest. The funders had no role in the design of the study; in the collection, analyses, or interpretation of data; in the writing of the manuscript, or in the decision to publish the results.

References

1. Schweitzer, G.; Maslen, E. *Magnetic Bearings: Theory, Design, and Application to Rotating Machinery*; Springer: Berlin, Germany, 2009.
2. Post, R.F. *Stability Issues in Ambient-Temperature Passive Magnetic Bearing Systems*; NASA STI/Recon Technical Report No. 3; NASA–Glenn Research Center: Cleveland, OH, USA, 2000.
3. Filatov, A.V.; Maslen, E.H. Passive magnetic bearing for flywheel energy storage systems. *IEEE Trans. Magn.* **2001**, *37*, 3913–3924. [CrossRef]
4. Detoni, J.G.; Impinna, F.; Tonoli, A.; Amati, N. Unified modelling of passive homopolar and heteropolar electrodynamic bearings. *J. Sound Vib.* **2012**, *331*, 4219–4232. [CrossRef]
5. Dumont, C.; Kluyskens, V.; Dehez, B. Linear State-Space Representation of Heteropolar Electrodynamic Bearings With Radial Magnetic Field. *IEEE Trans. Magn.* **2016**, *52*, 1–9. [CrossRef]

6. Amati, N.; Lepine, X.D.; Tonoli, A. Modeling of electrodynamic bearings. *J. Vib. Acoust.* **2008**. [CrossRef]
7. Amati, N.; Tonoli, A.; Zenerino, E.; Detoni, J.G.; Impinna, F. Design methodology of electrodynamic bearings. In Proceedings of the XXXVIII National Meeting of the Italian Society of Mechanical Engineers (AIAS), Torino, Italy, 8–11 September 2009.
8. Dumont, C.; Kluyskens, V.; Dehez, B. Yokeless radial electrodynamic bearing. *Math. Comput. Simul.* **2016**, *130*, 57–69. [CrossRef]
9. Dumont, C.; Kluyskens, V.; Dehez, B. Impact of the Yoke Material on the Performance of Wounded Electrodynamic Bearings. In Proceedings of the International Symposium on Magnetic Bearings, Linz, Austria, 11–14 August 2014.
10. Basore, P. Passive Stabilization of Flywheel Magnetic Bearings. Master's Thesis, Massachusetts Institute of Technology (MIT), Cambridge, MA, USA, 1978.
11. Filatov, A. Null-E Magnetic Bearings. Ph.D. Thesis, University of Virginia, Charlottesville, VA, USA, 2002.
12. Detoni, J.G.; Impinna, F.; Amati, N.; Tonoli, A. Rotational power loss on a rotor radially supported by electrodynamic passive magnetic bearings. In Proceedings of the 5th World Tribology Congress, Torino, Italy, 8–13 September 2013; pp. 1–4.
13. Kluyskens, V.; Dehez, B. Dynamical Electromechanical Model for Magnetic Bearings Subject to Eddy Currents. *IEEE Trans. Magn.* **2013**, *49*, 1444–1452. [CrossRef]
14. Tonoli, A.; Amati, N.; Impinna, F.; Detoni, J.G. A solution for the stabilization of electrodynamic bearings: Modeling and experimental validation. *J. Vib. Acoust.* **2011**, *133*, 021004. [CrossRef]
15. Davey, K.; Filatov, A.; Thompson, R. Design and analysis of passive homopolar flux bearings. *IEEE Trans. Magn.* **2005**, *41*, 1169–1175. [CrossRef]
16. Storm, M.E. The Application of Electrodynamic Levitation in Magnetic Bearings. Ph.D. Thesis, North-West University, Vanderbijlpark, South Africa, 2006.
17. Impinna, F.; Detoni, J.G.; Amati, N.; Tonoli, A. Passive Magnetic Levitation of Rotors on Axial Electrodynamic Bearings. *IEEE Trans. Magn.* **2013**, *49*, 599–608. [CrossRef]
18. Van Verdeghem, J.; Dumont, C.; Kluyskens, V.; Dehez, B. Linear State-Space Representation of the Axial Dynamics of Electrodynamic Thrust Bearings. *IEEE Trans. Magn.* **2016**, *52*, 1–12. [CrossRef]
19. Van Verdeghem, J.; Kluyskens, V.; Dehez, B. Experimental Validation and Characterisation of a Passively Levitated Electrodynamic Thrust Self-Bearing Motor. In Proceedings of the 2018 21st International Conference on Electrical Machines and Systems (ICEMS), Jeju, Korea, 7–10 October 2018; pp. 1863–1869.
20. Van Verdeghem, J.; Lefebvre, M.; Kluyskens, V.; Dehez, B. Dynamical Modelling of Passively Levitated Electrodynamic Thrust Self-Bearing Machines. *IEEE Trans. Ind. Appl.* **2019**. [CrossRef]
21. Amati, N.; Detoni, J.; Impinna, F.; Tonoli, A. Axial and radial dynamics of rotors on electrodynamic bearings. In *10th International Conference on Vibrations in Rotating Machinery*; IMechE, Ed.; Woodhead Publishing: Cambridge, UK, 2012; pp. 113–122.
22. Van Verdeghem, J.; Kluyskens, V.; Dehez, B. Five degrees of freedom linear state-space representation of electrodynamic thrust bearings. *J. Sound Vib.* **2017**, *405*, 48–67. [CrossRef]
23. Ebrahimi, B.; Khamesee, M.B.; Golnaraghi, F. Permanent magnet configuration in design of an eddy current damper. *Microsyst. Technol.* **2008**, *16*, 19. [CrossRef]

© 2019 by the authors. Licensee MDPI, Basel, Switzerland. This article is an open access article distributed under the terms and conditions of the Creative Commons Attribution (CC BY) license (http://creativecommons.org/licenses/by/4.0/).

MDPI
St. Alban-Anlage 66
4052 Basel
Switzerland
Tel. +41 61 683 77 34
Fax +41 61 302 89 18
www.mdpi.com

Actuators Editorial Office
E-mail: actuators@mdpi.com
www.mdpi.com/journal/actuators

www.ingramcontent.com/pod-product-compliance
Lightning Source LLC
LaVergne TN
LVHW070611100526
838202LV00012B/619